D0374894

Samuel Johnson

AND THE NEW SCIENCE

Samuel Johnson
AND THE NEW SCIENCE

RICHARD B. SCHWARTZ

The University of Wisconsin Press

MADISON, MILWAUKEE, AND LONDON

Published 1971
The University of Wisconsin Press
Box 1379, Madison, Wisconsin 53701

The University of Wisconsin Press, Ltd.
70 Great Russell Street, London WC1B 3BY

First printing

Printed in the United States of America
Heritage Printers, Inc., Charlotte, North Carolina

ISBN 0—299—06010—1; LC 76—161089

FOR
Judith and Jonathan

Contents

Acknowledgments

The wealth of recent Johnsonian scholarship is both impressive and heartening. My large debts to the work of Johnson's modern students will be apparent throughout the following pages. I am specifically indebted to the libraries of the University of Illinois and the University of Wisconsin for gracious and continuous support during the various phases of my work. In the procurement of materials, I have been aided by the British Museum, the Bodleian, the Boston Public Library, the University of Pennsylvania Library, the Library of Congress, the Yale University Library, and the Ramapo-Catskill Library System of the state of New York. Professors Robert Kargon, John Dussinger, and particularly Robert Haig provided invaluable help during the inception and early stages of the project. More recently, my colleagues, Professors Phillip Harth, Eric Rothstein, and Howard D. Weinbrot have provided advice, encouragement, and the wealth of their experience. Chapter IV appeared in abbreviated form in *Studies in Burke and His Time*; part of Chapter III appeared in the *Journal of English and*

Germanic Philology. My thanks are due to the editors of those journals for permission to reprint. My work was completed with the aid of a grant from the National Endowment for the Humanities, to which I express my great thanks. The greatest problems attending literary studies are frequently of a non-academic nature. Because of the untiring help of my wife Judith they did not arise.

Madison, Wisconsin   RICHARD B. SCHWARTZ
September, 1970

Samuel Johnson
AND THE NEW SCIENCE

ABBREVIATIONS

Adventurer; Idler	*Samuel Johnson: The Idler and The Adventurer*, ed. W. J. Bate, John M. Bullitt, and L. F. Powell (New Haven, 1963).
Journey	*Samuel Johnson: A Journey to the Western Islands of Scotland and James Boswell: The Journal of a Tour to the Hebrides with Samuel Johnson, LL. D.*, ed. Allan Wendt (Boston, 1965).
Letters	*The Letters of Samuel Johnson: With Mrs. Thrale's Genuine Letters to Him*, ed. R. W. Chapman, 3 vols. (Oxford, 1952).
Life	*Boswell's Life of Johnson*, ed. G. B. Hill, revised and enlarged by L. F. Powell, 6 vols. (Oxford, 1934, 1950).
Med. Dict.	R. James, M.D., *A Medicinal Dictionary* . . . , 3 vols. (London, 1743, 1745).
Prefaces & Dedications	*Samuel Johnson's Prefaces & Dedications*, ed. Allen T. Hazen (New Haven, 1937).
Tour	*Boswell's Journal of a Tour to the Hebrides with Samuel Johnson, LL. D., 1773, Edited from the Original Manuscript*, ed. Frederick A. Pottle and Charles H. Bennett (New York, 1961).
1825 *Works*	*The Works of Samuel Johnson, LL. D.*, 9 vols. (Oxford, 1825).

Introduction

In an article that has already generated considerable discussion and controversy, Donald Greene has advised the jettisoning of such misleading labels for eighteenth-century English literature as "Augustan" and "Neoclassic," and suggested that the phrases "Enlightenment" and "Age of Reason" be removed with equal haste. The picture of the age implicit in such descriptive terms simply does not fit the facts. Of far greater importance than the body of notions associated with the old categories are the age's anti-rationalist empiricism and a cluster of moral and theological beliefs best described as Augustinian. These, Greene argues, are due considerable attention in any attempt to describe the thought of the age, and their importance *can* be demonstrated in the case of individual writers.[1] His point is purposely overstated. After all, no self-

1. "Augustinianism and Empiricism: A Note on Eighteenth-Century English Intellectual History," *Eighteenth-Century Studies,* 1 (Fall 1967), 33–68. See also Greene's *The Age of Exuberance: Backgrounds to Eighteenth-Century English Literature* (New York, 1970), pp. 89–126. A useful

respecting lecturer omits a passing reference to the *Essay Concerning Human Understanding* and *Principia* as formative influences on the period, and the most superficial literary historian is certain to take note in some fashion of the Augustinian legacy in the eighteenth century and contrast the fallen, limited eighteenth-century world with the Romantic ethos of infinite possibility. This in no way lessens the force of Greene's argument, for the question at hand is one of stress. With regard to empiricism, for example, Greene is well aware of the volume of material dealing with Locke and Locke's influence, as well as the continually growing body of commentary on the relationship between science, philosophy, and literature in the period. He is also aware, however, that such material has not been sufficiently utilized to effect a substantial alteration in overviews of that period.

Whether a major revaluation of the age is in order is open to question, but the importance of the English tradition of science and scientific ideology—to which the empirical methodology is central and in which much later philosophy found its inspiration —should be as evident to us as it was to Voltaire. French admiration for English science at this time is well known, but English praise has, for reasons I will later suggest, been often overlooked. The views of the major writer of the period, my concern in this study, have not been considered in sufficient detail, and in some cases have been totally misunderstood.

The course of modern scholarship in the case of Johnson is particularly instructive. In his pioneer study, *The New Science and English Literature in the Classical Period*, published in 1913, Carson S. Duncan noted Johnson's antipathy to the satirists of

---

principle in the study of the history of ideas is that the examination of contemporary textbooks provides a representative index to the period's attitudes. The best recent textbook for eighteenth-century literature is the Tillotson, Fussell, Waingrow anthology (*Eighteenth-Century English Literature* [New York, 1969]) where one finds (pp. 1-5) a reflection of Greene's point of view concerning the fatuity of most labels ("The Enlightenment," "The Age of Reason," "The Neoclassical Period," "The Augustan Age") and the commanding importance for the period of English empiricism.

science and called attention to Johnson's biography of the Dutch physician Hermann Boerhaave. The book was privately printed and did not receive the attention it deserved. Statements such as the following continued to appear:

[Johnson] had, unfortunately, little or no interest in natural science, no scientific curiosity, and lacked entirely the high type of imagination that causes awesome wonder at that great machine, the universe.[2]

And yet it is in his "Ramblers" that Johnson most clearly reveals himself and his views—his dislike of science and scientific speculation, his love of sound morals and sound sense.[3]

The circle of Johnson's immediate friends and admirers held the most eminent names in English letters, art, politics, and the contemporary stage. . . . And it is curious, reflecting upon this, that while the sage omniscience of Johnson is appealed to at all turns, nobody within that circle seems to have been conscious of the importance to the history of thought of the contemporary school of English empirical philosophy . . . .[4]

In conclusion, what was Johnson's attitude to science in general? Perhaps it is surprising that in the eighteenth century science should have been a sufficiently demarcated and unified body of ideas to evoke reflections upon its value, still more that Johnson should give the matter any thought. In his view, science could offer pleasing diversions, but ought not to be taken more seriously . . . .[5]

Approximately forty years after the appearance of Duncan's book, Jean Hagstrum once again pointed out Johnson's sympathies for science and also his Modern posture with regard to the Battle of the Books and the Ancients-Moderns controversy. In a tone of fresh discovery he noted the importance of the *Life*

2. Charles W. Burr, "Some Medical Words in Johnson's Dictionary," *Annals of Medical History*, 9 (June 1927), 185.

3. Charles Whibley, "Samuel Johnson: Man of Letters," *Blackwood's Magazine*, 221 (May 1927), 669.

4. Harold Williams, "Dr. Johnson's Favourite Pursuits," *Nineteenth Century*, 123 (May 1938), 620.

5. Russell Brain, "Doctor Johnson on Science," in *Some Reflections on Genius and other Essays* (London, 1960), p. 46.

of Boerhaave. Of course, W. K. Wimsatt's *Philosophic Words* had appeared several years earlier and MacLean's famous study of Locke had noted the attraction Locke's principles held for Johnson. Since that time a number of individuals have stressed Johnson's empiricism, skepticism, and debts to Locke, and anyone who has inspected the works of Greene, Robert Voitle, Maurice Quinlan, and Paul Alkon, to name only a few, must be struck by the near universal acceptance of the fact that Johnson is a Modern, a Lockean empiricist, a man aware of and in sympathy with the methodology traceable to the new science.

For all this, one still encounters judgments such as Paul Fussell's, that Johnson's assessment of scientific knowledge is "low,"[6] and a recent commentator, surveying the interaction between science and literature in the eighteenth century, wrote that "Samuel Johnson liked chemistry and added much science to the 'philosophic words' of his dictionary, but in his essays and criticism he seldom showed more than a passing interest in the subject. The four *Rambler* papers that comment on science (82, 83, 177, 199), for example, consist mainly of conventional satire on the virtuoso."[7] In his recent introduction to Johnson and Boswell, the late E. L. McAdam wrote of the *Life of Boerhaave* that it "has little interest for the modern reader."[8]

It is difficult to explain this disparity between settled consensus and outworn opinion. It can be traced, I think, to several causes, one of which Greene has often noted, the fact that Johnson is simply not studied with the rigor which a major figure demands and deserves. Also, the uncertain canon, the absence until recent years of a standard edition, and the awesome body of antiquarian, critical, and scholarly commentary, have created a kind of labyrinth open only to the thoroughgoing specialist. This, along with the tendency, based on some remarks in the periodical essays and *Life of Milton*, to lump Johnson with the

6. *The Rhetorical World of Augustan Humanism: Ethics and Imagery from Swift to Burke* (Oxford, 1965), p. 17.

7. William Powell Jones, *The Rhetoric of Science: A Study of Scientific Ideas and Imagery in Eighteenth-Century English Poetry* (London, 1966), p. 179.

8. *Johnson & Boswell: A Survey of Their Writings* (Boston, 1969), p. 9.

Restoration and eighteenth-century satirists of science, may partially account for the fact that there is still considerable misunderstanding with regard to his attitudes toward the new science. "Locke," Voltaire wrote, "unfolded to man the nature of human reason as a fine anatomist explains the powers of the body. Throughout his work he makes use of the torch of science."[9] Johnson's empiricism is intimately bound up with his knowledge of and attitudes toward the new philosophy. If we are to see that empiricism and its attendant skepticism in proper perspective we must refer, as Voltaire does, to the larger context.

There is a clear relation, for example, between his approval of the methodology of science and his constant demand for veracity and exactitude. Of course, his detestation of liars and the manner in which they destroy that confidence on which happiness and organized society depend, is a fundamental cause of his desire for accuracy, but as he indicated to Mrs. Thrale, "It is more from carelessness about truth than from intentional lying, that there is so much falsehood in the world."[10] A goodly portion of that carelessness may be attributed, in Johnson's judgment, to a pre-scientific intellectual barbarism, the credulity and disregard for detail and firsthand observation which he assails in the *Journey to the Western Islands of Scotland*. It is the Johnson who has grown accustomed to the importance of observation, experimentation, and verification who objects to the witchery of *Macbeth*:

A poet who should now make the whole action of his tragedy depend upon enchantment, and produce the chief events by the assistance of supernatural agents, would be censured as transgressing the bounds of probability, be banished from the theatre to the nursery, and condemned to write fairy tales instead of tragedies . . . .[11]

9. *Philosophical Letters*, trans. Ernest Dilworth (New York, 1961), p. 54.
10. *Life*, III, 228–29. Cf. *Life*, II, 434; Hester Lynch Piozzi, *Anecdotes of the Late Samuel Johnson, LL.D.*, ed. G. B. Hill in *Johnsonian Miscellanies* (Oxford, 1897), I, 225, 241, 243, 297, 347–48; J. D. Fleeman, "Johnson and the Truth," in *Johnsonian Studies*, ed. Magdi Wahba (Cairo, 1962), pp. 109–113.
11. *Johnson on Shakespeare*, ed. Arthur Sherbo, 2 vols. (New Haven, 1968), VIII, 752.

Thus, in discussing Johnson's attitudes, we are of necessity also discussing an approach to experience, an intellectual temper or cast of mind which helps to explain or enhance our present understanding of much of his work.

Johnson's views have received attention from various scholars. John J. Brown's early study, "Samuel Johnson and Eighteenth-Century Science,"[12] deals with the extent and depth of his scientific knowledge. Though helpful in matters of biography, little attention is given to Johnson's attitudes. Wimsatt has been primarily concerned with biography, linguistic history, and rhetoric, and has deepened our understanding of the Johnsonian style; but, again, he does not deal with attitudes in sufficient detail. Hagstrum has, I feel, suggested the most useful approaches to the subject, but since he is primarily interested in the relevance of Johnson's attitudes to his literary criticism, he devotes, quite understandably, less attention to the attitudes than I feel they deserve. In his dissertation, "Samuel Johnson and the New Philosophy," Stephen O. Mitchell is chiefly concerned with Johnson's cosmology.[13] The issues discussed in my study are, in general, peripheral in Mitchell's. His conclusions concerning Johnson's "rejection of virtuosity" and opposition to the scientific ethos are not, I believe, supported by the available evidence, and writings long attributed to Johnson, which shed considerable light on his attitudes, are not taken into account. Besides these studies, a large number of articles, many difficult of access, have dealt in one way or another with Johnson's connections with science.

In view of the lingering uncertainty concerning Johnson's reaction to the new science and the demonstrated importance of English empiricism in his methodology of cognition, a consideration of his views of science is, I feel, justified. Further demonstration of the breadth and depth of his interests is always useful, but my main focus is not Johnson's scientific erudition. He is a commentator, not an experimenter. No scientific discoveries

12. "Samuel Johnson and Eighteenth-Century Science," Diss. Yale 1943.
13. "Samuel Johnson and the New Philosophy," Diss. Indiana, 1961.

can be associated with his personal chemical diversions and his attempts to deal with technical scientific material are extremely rare. His tendency is to view scientific problems and accomplishments in the context of general experience, to treat science's position within a framework of larger patterns. Of course, the scientists themselves did this also, so that Johnson's remarks belong to the tradition of scientific ideology and polemic, rather than the more technical line of discourse. Thus, my encapsulation of Johnson's scientific experiences and "education" in the second chapter is intended to demonstrate the fact that his familiarity with scientific developments is sufficiently extensive to justify his role as commentator. In the second chapter I also attempt to provide an indication of the nature and number of the works engendered by scientific interests, scientific learning, and often the catalyst of penury. My primary interest, however, lies in his views of the Baconian scientific tradition, the satirists who opposed it, and the divines, scientists, and laymen who defended it. My stress, I hope, parallels Johnson's and the organization of my study is to a large extent dictated by the nature of his commentary.

One problem which everyone faces in writing on Johnson is the question of audience, a difficulty resulting from what Bertrand Bronson terms "the double tradition." Inevitably, one finds himself preaching simultaneously to the converted and the infidel. The "popular" Johnson refuses to die; the infidel—nurtured by nineteenth-century commentators and strengthened by a careful avoidance of the reading of Johnson's works—refuses to let the genuine Johnson live. One of my purposes in this study has been to marshal sufficient evidence to demonstrate Johnson's sympathies for English science, his "Modernism," openness, curiosity, and tendency to justify and explain rather than attack, so that the tired judgment that he is fundamentally opposed to the new philosophy might finally be laid to rest. For some I will be repositing a straw man, for others spreading virulent heresy. If for some I belabor the obvious and for others take for granted what calls for rigorous demonstration, I ask indulgence. The fact that some of the material I cite has only recently been attributed to Johnson and remains unfamiliar to many students of the eighteenth century seems to me to justify more extensive quotation than is perhaps usual. Further, I have always assumed that the clearest articulation of Johnson's views comes from Johnson himself. Whenever possible I have allowed him to speak.

CHAPTER I

An Age of Science

Early in the Essay of Dramatick Poesie, Crites discusses the "universal genius" peculiar to each age, "which inclines those that live in it to some particular studies . . .":

> Is it not evident, in these last hundred years (when the study of philosophy has been the business of all the Virtuosi in Christendom), that almost a new Nature has been revealed to us?—that more errors of the school have been detected, more useful experiments in philosophy have been made, more noble secrets in optics, medicine, anatomy, astronomy, discovered, than in all those credulous and doting ages from Aristotle to us?—so true is it, that nothing spreads more fast than science, when rightly and generally cultivated.[1]

No denials are made, for Crites, at least in these matters, is on safe ground. The great importance of a proper methodology, implicit in the final sentence, is the continual concern of Bacon and his successors. The audacity which debunks scientific myth

---

1. W. P. Ker, ed., *Essays of John Dryden,* I (Oxford, 1900), 36–37.

11

and topples the tautological, qualitative science of peripateticism and its scholastic adherents is a hallmark of the new philosophy. Though the satirists and defenders of the past frequently questioned the usefulness of the experimental philosophy—the *real* Philosophy as Petty and others termed it[2]—that usefulness was judged, as Crites feels, clearly demonstrable. And since Whitehead's "century of genius" is informed by the work of such men as Bacon, Hariot, Gilbert, Kepler, Galileo, Harvey, Fabricius, Huyghens, Gassendi, Hobbes, Descartes, Leibniz, and preeminently Newton, little rejoinder is possible to Crites' claim that science had advanced on a broad front and permanently altered man's conception of the external world. The successive revolutions in chemistry and the natural sciences were yet to be effected, and medicine had hardly advanced to the degree which Crites suggests, but the palpable changes in human thought which the seventeenth-century scientists had accomplished were, by 1668, virtually irrevocable. The century was an age of science, no less than the succeeding period in which the work of the seventeenth century would be integrated and extended.

The most frequently cited index to the temper of the period, and to the nature of the scientific pursuits which formed a considerable portion of the influence on that temper, is the extent to which scientific study, even in especially humble forms, was conducted by the philosophical amateur. The levels of "virtuosity" that may be discerned are extremely varied, but the breadth of scientific interest and the sheer number of individuals affected is undeniably great. Shakespearian London could boast the accomplishments of a man like Henry Percy, the "Wizard Earl," who patronized an entire school of natural philosophers, refused to travel without his chemical equipment, and who, upon being sent to the Tower (on very flimsy evidence of com-

2. Richard Foster Jones, *Ancients and Moderns: A Study of the Rise of the Scientific Movement in Seventeenth-Century England*, 2nd ed. (St. Louis, 1961), p. 293, n. 7. Cf. Hooke, *Micrographia* (London, 1665), sig. a2ʳ. Sprat, in his *History of the Royal Society*, ed. Jackson I. Cope and Harold Whitmore Jones (St. Louis, 1966), speaks constantly of "real knowledge" and "true philosophy."

plicity in the Gunpowder Plot) took with him crucibles, retorts, alembics, charts, globes, and even, it is said, human skeletons. By 1614, his quarters in the Martin Tower were so cluttered with scientific apparatus and paraphernalia that he was forced to rent from Lord Carew the neighboring Brick Tower. When one offer of pardon came in 1617 he refused it, preferring to remain in his prison laboratory.[3]

The number of such individuals increased after the Restoration, and the virtuosi, who always offered encouragement and sometimes financial support, constituted a willing audience for the serious scientist. It is doubtful that the Royal Society could have progressed to the extent that it did without their willing and continual aid.[4] Though they often incurred the wrath of satiric observers, and though the Royal Society was to be hampered by the long-standing high proportion of non-scientific membership, a problem not alleviated until the nineteenth century, the broad reception of science, which the virtuosi facilitated, was an absolute necessity during the time at which opposition began to appear.

The number of literary men who showed interest in the new science has often been remarked. Dryden, Waller, Denham, Cowley, Pepys, and Evelyn were all closely associated with the Royal Society. Cowley's ode to the Royal Society, included with Sprat's history, is an important document in the tradition of scientific ideology and polemic. Before the Restoration, at a point when Royalist hopes had all but faded, Evelyn contemplated the founding of a kind of scientific retreat to which he and his wife could retire, and after the founding of the Royal Society became an active and productive member. In 1687 Pepys, as President of the Society, licensed the publication of Newton's *Principia*, and in a delightful passage in his diary (26 May 1667) indicated the manner in which scientific tinkering could embellish his other interests:

3. Robert Hugh Kargon, *Atomism in England From Hariot to Newton* (Oxford, 1966), pp. 5–17.

4. Dorothy Stimson, *Scientists and Amateurs: A History of the Royal Society* (New York, 1948), pp. 27, 36, 115.

Much against my will staid out the whole church . . . but I did entertain myself with my perspective glass up and down the church, by which I had the great pleasure of seeing and gazing at a great many very fine women; and what with that, and sleeping, I passed away the time till sermon was done.[5]

Even Charles, who, with the court poets, could wittily question the accomplishments of the new philosophers, prided himself on his scientific learning and had his own laboratory at Whitehall where he performed dissections of human bodies.

When we turn to Johnson's century, the number of important writers with an amateur or serious interest in science multiplies. Pope was an inveterate collector of curiosities as well as a student of Newtonian astronomy and Cartesian cosmogony.[6] Even Swift, so vocal a critic of the new science, presented a microscope to Stella and enjoyed, as Pope wrote to Arbuthnot, "an Orbicular Glass, which by Contraction of the Solar Beams into a proper Focus, doth burn, singe, or speckle white, or printed Paper, in curious little Holes, or various figures."[7] Cowley, Garth, Blackmore, and Akenside, whose lives Johnson would write, were all, like Goldsmith and Smollett, physicians. Gray spent the majority of his final decade studying natural history, while Adam Smith and Gibbon attended William Hunter's course in anatomy. Gibbon also took lessons in chemistry and interested himself in natural history; in his *Essays on Philosophical Subjects*, published posthumously in 1795, Smith left a considerable fragment on the history of astronomy. The same pattern is mirrored on the continent. Besides the formidable importance of Voltaire's popularizations of Newton and English science, Diderot's writings include a work on the elements of physiology and Rousseau prepared a treatise on the fundamental

5. Cited by Marjorie Nicolson, *Pepys' "Diary" and the New Science* (Charlottesville, 1965), pp. 22–23.

6. For a discussion of Pope's reaction to the new philosophy, see Marjorie Nicolson and G. S. Rousseau, *"This Long Disease, My Life": Alexander Pope and the Sciences* (Princeton, 1968).

7. *The Correspondence of Alexander Pope*, ed. George Sherburn, i (Oxford, 1956), 234. The letter is dated 11 July 1714.

laws of chemistry. Montesquieu treated physical and physiological problems; d'Alembert (like Pascal) was a brilliant mathematician and Goethe's career received much of its luster from his genuine scientific contributions to botany, optics, and comparative anatomy.

It would be misleading to assume that science was a safely compartmentalized activity, to be studied in solitude by the serious or toyed with by the amateur when idle leisure would permit. Many of the implications of scientific learning were already being realized, and scientific progress or retrogression hinged on the interrelations between the new philosophy and other, both private and public, areas of human experience. For example, science could be, and was, associated with Puritan proposals for revolutionary educational reform, associations which were, as might be expected, played down after the Interregnum so that the feelings of such bodies as the universities could be assuaged. Bacon's delineation of the necessary intellectual equipment for studying science, and Locke's epistemology, which also to an important extent put all men on an equal footing, had definite and by now often noted affiliations with the democratic spirit. At worse, science might be termed foul whiggery by more conservative spirits, and Newton, as Master of the Mint during the troubling episode of Wood's coinage, was, after all, a Whig politician. With the drawing of sharp ideological lines, the scientific future could be viewed with heady enthusiasm or auguries of doom, a situation intensified by the anti-scientific impulse motivating satirists whose concerns could be articulated in works of literary genius.

The possibility of science's falling afoul of established religion was a continual source of anxiety and cause of considerable rationalization. Leonardo and Roger Bacon had been silenced by the Church's power. Copernicanism was opposed from the beginning by Luther and the reformers. Servetus, a fellow student of Vesalius, who had written on the circulation of the blood, had been burned at the stake by Calvin for his unitarian ideas. The cases of Bruno, Campanella, and Galileo were still fresh in contemporary memory. Thomas Hariot, perhaps the greatest scientist of Shakespearian England, was, despite the lack of real evidence, arrested and imprisoned under suspicion of atheism.[8] Hobbes's flagrant unorthodoxy inhibited the open study or espousal of atomism, which already suffered from the long-standing stigma of atheism.[9] Even *Religio Medici*, which Johnson would defend in his biography of Sir Thomas Browne, had been placed on the Roman Catholic Index.

The scientists, though they wished to set scientific studies on a proper footing by revealing and discarding the weaknesses of the science of antiquity and the middle ages, did not wish to disturb the religious beliefs which constituted one of the most cherished segments of English thought. On the contrary, science should enforce religion, not undermine it, and the discoveries made possible by, for example, the telescope and microscope, should underline the fact that the heavens as well as the microscopic worlds declare the glory of God and the divine wisdom that penetrates all creation. Thus, Boyle's famous endowment of a series of lectures for the purpose of refuting atheism, the number of bishops and other clergymen who were active in the Royal Society (which Glanvill and Sprat were quick to point out), and the grand example of Newton's piety were all proof of the scientists' good faith. The attempts to join scientific discovery with theological principle are omnipresent. Walter Charleton wrote a treatise refuting atheism. Evelyn, in an essay on Lucretius, enlisted atomism in the defense of religion. Richard Bentley, Henry More, John Ray, William Derham, Samuel Clarke, William Whiston, Henry Baker, Sir Richard Blackmore, and especially Robert Boyle, to mention only a few, took up the scientific cudgel in the defense of religion, in general by copious argument for the existence of God on the basis of the principle of design in both the physical universe and the human body.

8. Kargon, *Atomism . . . From Hariot to Newton*, pp. 27–29.

9. Ibid., pp. 60–62, 76, 89. For a full treatment of unfavorable contemporary reactions to Hobbes, see Samuel I. Mintz, *The Hunting of Leviathan: Seventeenth-Century Reactions to the Materialism and Moral Philosophy of Thomas Hobbes* (Cambridge, 1962).

The important relationship between science and philosophy in the late seventeenth and eighteenth centuries is well known. The grand preoccupation with epistemology, running in an unbroken line through Locke, Berkeley, and Hume, is inexplicable when separated from the larger context of scientific methodology. Henry More, who corresponded with Descartes and van Helmont, was, like Locke, a Fellow of the Royal Society. Hobbes, mainly remembered today for his philosophy and political theory, was one of the chief mechanical philosophers at mid-century. Nearly every thinker was affected in some way by the new philosophy, and specific influence, such as, for example, that of the Newtonian theory of vibrations on Hartley, is not uncommon.

More interesting for our purposes, however, is the omnipresent relation between science and literature. The manifold connections of science—apparent to a contemporary mind like Swift's or Johnson's—constitute an extremely complex web of associations that vibrates at the touch. This is why we so often find literary figures dealing with scientific or quasi-scientific matters, and why both the recognition of the use of scientific rhetoric and an awareness of the writers' preoccupation with the implications of the new science are fundamental to an understanding of Restoration and eighteenth-century English literature.

Literary figures whose reaction to the new science was favorable were particularly influenced by the authors of popular, physico-theological compendia. Such writers as Boyle and Ray were used by physico-theologians like William Derham, who in turn were picked up by poet and prose writer alike. William Powell Jones has demonstrated the ubiquitous effect of physico-theology on both major and minor literary figures.[10] Concepts such as the following, common in the literature of the period,

10. *The Rhetoric of Science* (London, 1966), pp. 20–21, 28–30, et passim. See also Jones's articles, "Science in Biblical Paraphrases in Eighteenth-Century England," *PMLA*, 74 (March 1959), 41–51; "The Idea of the Limitations of Science from Prior to Blake," *Studies in English Literature*, 1 (Summer 1961), 97–114.

can be traced to this immediate influence: (1) the sense of an orderly universe set in motion by a divine creator, dramatized both on the grand and miniature scales by the respective developments in telescopic and microscopic studies; (2) the universe's plenitude, expressed through the dominant metaphors of the great chain of being and the book of nature which lies open to all; (3) the importance of Providence which reconciles man to such apparent evils as earthquakes, hurricanes, and inequalities of climate, all of which are necessary to fill out the divine scheme of plenitude; (4) the limitations of science—science's inability to answer the fundamental questions as to the cause behind observable effects. Poets or essayists, depending on their own intellectual propensities and the degree to which they were influenced by physico-theology, generally developed these basic notions. The influence on a poem like the *Essay on Man* should be evident. The fourth notion is both common and important. Pope makes the point, as does Cowper:

> But never yet did philosophic tube,
> That brings the planets home into the eye
> Of observation, and discovers, else
> Not visible, his family of worlds,
> Discover him that rules them; such a veil
> Hangs over mortal eyes, blind from the birth,
> And dark in things divine.
>
> (*The Task*, III, 229–35)

Cowper is not necessarily averse to science as such, but realizes and stresses the fact that science does not in and of itself hold the answers to man's most basic questions. Piety must be added, and when it is, as in the case of Newton, Cowper's praise is profuse:

> Philosophy, baptiz'd
> In the pure fountain of eternal love,
> Has eyes indeed; and, viewing all she sees
> As meant to indicate a God to man,
> Gives *him* his praise, and forfeits not her own.
> Learning has borne such fruit in other days

On all her branches: piety has found
Friends in the friends of science, and true pray'r
Has flow'd from lips wet with Castalian dews.
Such was thy wisdom, Newton, childlike sage!
Sagacious reader of the works of God,
And in his word sagacious.

(III, 243–54)

Although many poets and essayists explored the ways in which astronomy, microscopy, human physiology, botany, and even the animals and insects provided evidence of the wisdom of God, science was put to other uses than the proof of the existence of God or the description of the beauties of His creation. Besides didactic poetry like the *Essay on Man*, hexaemeral poetry like Blackmore's *Creation*, and encyclopedic descriptive poetry like the *Seasons*, all of which were indebted to physico-theology, there are works of other types which find their material or inspiration in science. These include voyages to the moon, often influenced by Kepler's famous *Somnium*,[11] poems describing the cosmic voyage of the scientist as he explores the universe after death (a genre to which Anna Williams'/Samuel Johnson's "On the Death of Stephen Grey, F.R.S." belongs), or the use, in critical discussions, of such notions as Newtonian light and color, the joining of science with aesthetics.[12] The available information on science, coupled with the fact that it could be, and was, put to very diverse uses underlines the fact that no schematic outline can hope to summarize faithfully the reaction to science and use of scientific material by eighteenth-century literary figures.

In general we may say that there was a favorable reception of

11. See Marjorie Nicolson, "Kepler, the *Somnium*, and John Donne," in *Science and Imagination* (Ithaca, 1956), pp. 58–79.

12. On science and aesthetics, see Miss Nicolson's *Newton Demands the Muse: Newton's "Opticks" and the Eighteenth-Century Poets* (Princeton, 1946); and her masterful *Mountain Gloom and Mountain Glory: The Development of the Aesthetics of the Infinite* (Ithaca, 1959). Cf. William Powell Jones, "Newton Further Demands the Muse," *Studies in English Literature*, 3 (Summer 1963), 287–306.

science, the major representative of which is Thomson, but today the effects of science are remembered chiefly for the attacks they called forth from the satirists of the Restoration and the reign of Queen Anne. In terms of sheer poetic bulk, the age is sympathetic to science, but the works of the satirists clearly preponderate in quality. *The Battle of the Books*, which marks the most important literary episode in the English Ancients-Moderns controversy, as well as Swift's attacks on science in the *Tale of a Tub* and *Gulliver's Travels* have outlived the popular praise of science in the minor verse of the period. Modern readers turn to the Scriblerus *Memoirs, Three Hours after Marriage*, and Pope's numerous comments in the *Essay on Man* and *Dunciad* as well as his scattered remarks on the petulant geologist, John Woodward, far more quickly than they pick up a work like Blackmore's *Creation*.

To say that there are two clear-cut eighteenth-century reactions to science, favorable and unfavorable, is to sacrifice truth for convenience. It is possible to determine some common ground in the works of science's admirers and it is easy to list a large number of attacks on science from, for example, Butler, Shadwell, Aphra Behn, Swift, Pope, Arbuthnot, King, Wolcot, and Blake. But, as in the case of the physico-theological writers, we cannot lump the satirists together in any useful way. The variety of their individual grievances and differences in their modes of attack will not permit any neat summaries. The satire on science in Shadwell's *Virtuoso* or Pope, Gay, and Arbuthnot's *Three Hours after Marriage* represents the use of science for topical humor. The type of science attacked can be silly tinkering, as in *The Virtuoso*, pedantry, as in the Scriblerus *Memoirs*, or the dangerous pursuit that may end in human pride, as in the *Essay on Man*. The meanings and exploitations of "philosophy" are protean. Swift can lament modernism in general and science in particular while Arbuthnot attacks an overzealous regard for antiquity. Among the satirists are men like Pope who can praise some individuals and denigrate others. Newton receives his oft-quoted epitaph, Woodward continual criticism. In the pages of the *Tatler, Spectator*, and *Guardian*, criticism of virtuosi exists side by side with discussions of the great chain of being and the spacious firmament on high.[13] As with the scientific encomiasts, the placing of satirists in camps, schools, or splinter sects can only be useful in the case of those writers whose works are marked by feeble imitation.

If we must summarize the reaction of literary figures to the new philosophy we must be content to say that some writers received with admiration science, its methods, conclusions, and religious implications, as well as the usefulness of scientific findings in the traditional forms of descriptive and didactic verse. Some writers accepted it with certain reservations; others found in it a source for satire and, with greater or less attachment to an anti-Modern humanism and fears of the ramifications of science in the fields of education, politics, or religion, attacked it in various ways. The importance of science for the period's literature is unmistakable and its influence can, to a limited degree, be traced, but the extent of that influence, and the ways in which it manifests itself must be determined by close examination of each eighteenth-century author.

Before turning to Johnson's experiences with, writings on, and attitudes toward the new science, it should be stressed that eighteenth century science, to a great extent, is English science. It is true that natural philosophy, like few other disciplines, knows no geographical or political boundaries. Foreign scientists have from the beginning been important Fellows of the Royal Society. Correspondence with Christiaan Huyghens was maintained in spite of the existence of a state of war between England and Holland. During the Revolutionary War, Franklin, a distinguished Fellow and recipient of the Royal Society's Copley Medal, instructed American ships not only to spare but to aid Captain James Cook, and Biot and Gay-Lussac were elected Fellows of the Society just before Waterloo.[14] Nevertheless,

13. Relevant papers in the Addison-Steele periodicals include *Tatler* 15, 111, 119, 216, 221, 236; *Spectator* 21, 111, 120, 121, 242, 262, 387, 393, 413, 420, 465, 519, 531, 543, 554, 565, 588, 626, 635; *Guardian* 24, 27, 35, 69, 70, 75, 103, 107, 112, 126, 156, 169, 175.

14. Sir Gerard Thornton, "International Bond of Science," in *The Royal Society Tercentenary* (London, 1961), pp. 62–69; Sprat, *History of the Royal Society*, pp. 62–63, 127–28.

British science was the world's model. Continental Anglomania at this time is common. In the words of a recent commentator, Paris "was the modern Athens, the preceptor of Europe, [but] it was the pupil as well. French philosophes were the great popularizers, transmitting in graceful language the discoveries of English natural philosophers and Dutch physicians. . . . The propagandists of the Enlightenment were French, but its patron saints and pioneers were British: Bacon, Newton, and Locke . . . ." "British empiricism transformed French rationalism . . . ."[15]

Voltaire, in extremely Johnsonian fashion, praises Newton in exorbitant terms and professes allegiance to men who hold "sway over men's minds by force of truth, not to those who make slaves by violent means . . . ."[16] The like of Newton "is not seen in ten centuries"; Bacon is "the father of experimental philosophy"; there has perhaps "never been a wiser, more orderly mind, or a logician more exact, than Mr. Locke"; Descartes was "led astray by that spirit of system which blinds the greatest of men"; Malebranche suffers from "sublime hallucinations."[17] These statements are tempered throughout the *Philosophical Letters*, and Voltaire does, for example, cite many of the shortcomings of the Royal Society, but his exuberance and admiration for things English is undeniable and his importance as a promulgator is great. For d'Alembert, who fully grasped the significance of English science, Newton "gave philosophy

15. Peter Gay, *The Enlightenment: An Interpretation,* i (New York, 1968), 11, 13.

16. *Philosophical Letters,* trans. Ernest Dilworth (New York, 1961), p. 46. Cf. Johnson's discussion of military projectors in *Adventurer* 99. The notion resounds throughout the period; see, for example, *Essay on Man,* iv, 230–36; Gray, *Elegy Written in a Country Church-Yard,* 67–68; Sprat, "Epistle Dedicatory" to his *History of the Royal Society,* par. 1: "For, to increase the Powers of all Mankind, and to free them from the bondage of Errors, is greater Glory than to enlarge *Empire,* or to put Chains on the necks of Conquer'd *Nations.*" Sir Andrew Freeport (*Spectator* 2) "will tell you that it is a stupid and barbarous Way to extend Dominion by Arms; for true Power is to be got by Arts and Industry." Vainglorious military figures abound in *Paradise Lost.* See, e.g., vi, 376–85; ix, 27–41; xi, 689–99; xii, 24–78. For other parallels, see Gay, *The Enlightenment: An Interpretation,* ii (New York, 1969), 50, 51, 123–24.

17. *Philosophical Letters,* pp. 46, 48, 52, 53.

a form which apparently it is to keep"; Locke "reduced metaphysics to what it really ought to be: the experimental physics of the soul . . ."; and Bacon, the "immortal Chancellor," should be placed at the head of those who "prepared from afar the light which gradually, by imperceptible degrees, would illuminate the world."[18]

English science received considerable publicity in France, not only from its own popularizers, but through the efforts of Dutch physicians such as Boerhaave, 'sGravesande, and Musschenbroek,[19] a particularly interesting link in the spread of scientific ideology, since Johnson asserted that he had not read Bacon until the preparation of the *Dictionary*. Yet, as early as 1739 when he published the first version of his biography of Boerhaave, he is in firm possession of the principles which he would defend and enlarge upon throughout his life. We cannot assume that this single project is the primary source of his scientific education. The search for unique sources, both in intellectual history and literary studies, nearly always guarantees falsification in the case of major writers, and Johnson himself indicated that his reading was broader and more intense in his early years than in his later. We can say, however, that if his early reading did not include Bacon and if his readings in natural philosophy were not yet extensive, he could certainly have been thoroughly initiated into the ideological mainstream through his work on Boerhaave.

Boerhaave's *De Comparando Certo in Physicis* (1715), which Johnson discusses, had a wide circulation in France. Boerhaave stresses the importance of Baconian methodology and attributes the contributions of Boyle, Halley, and Newton to that influence. Boerhaave was made a member of the Académie des Sciences in 1731; La Mettrie went to Leyden to study under him; Voltaire consulted him in 1737 concerning his personal health. In 1715 'sGravesande went to England as secretary to

18. *Preliminary Discourse to the Encyclopedia of Diderot*, trans. Richard N. Schwab (New York, 1963), pp. 81, 84, 74.

19. See Pierre Brunet, *Les Physiciens Hollandais et la Méthode Expérimentale en France au XVIIIᵉ Siècle* (Paris, 1926); Gay, *The Enlightenment: An Interpretation*, II, 135–37.

the embassy which was to felicitate George I on his accession. While there he met Newton and Desaguliers, became a member of the Royal Society, and later openly professed his admiration for Newtonian methodology. Musschenbroek, a student of Boerhaave and colleague of 'sGravesande, also worked with Newton and Desaguliers. In 1734 Réaumur eulogized him before the Académie des Sciences. The Abbé Nollet became a Fellow of the Royal Society in 1734 and later became associated with 'sGravesande, Musschenbroek, and Allamand in Holland. His influence on French experimental science was considerable. The works of the Dutch physicians did not always meet with approval in still-Cartesian France, but the experimental method of English science received publicity and an important degree of acceptance from their writings and personal contacts.

An interested eighteenth-century observer of the revolutions in scientific thought confronted two complementary traditions: the series of technical treatises and the attendant body of polemic and ideological statement. To a degree the two are inseparable and often coexist in the pages of single volumes. The need for justification, rationalization, and methodological excursuses was felt by nearly all; the temptation to generalize on the progress of modern learning, cite its palpable accomplishments, and tout its successful *modus operandi* was seldom resisted. More important, scientific developments were consistently linked with a number of issues and controversies which touched at its center the intellectual, religious, political, and social life of the country. Some possible ramifications, implications, and effects of science were sensed immediately, and "philosophic" writers did not hesitate to add their opinions to the swelling body of ideological commentary. It is to this tradition that Johnson belongs. His tendency to assess the functions and importance of science in the contexts of other human concerns is mirrored in the practices of the scientists themselves, and Johnson treats nearly every important issue which was bandied about in the prefaces of scientific works and the individual polemics, apologias, proposals, and diatribes spawned by Baconian science. He is a commentator, not a serious experimenter or professional "philosopher," but the role he assumes is justified by his personal scientific learning and immeasurably enhanced by his knowledge of and experience with the human issues explicitly related to scientific methods and advances. Whether as essayist, biographer, travel writer, poet, reviewer, or preface writer, the quality of his commentary on scientific matters often surpasses the writings of the scientists themselves, because of the reserves of intellectual sophistication, rhetorical skill, and personal experience on which he is able to draw. However, before turning to Johnson's commentary, we must sketch, in outline, the ideological tradition to which he contributed.

In indicating the number of issues both central and peripheral to the Ancients-Moderns controversy, R. F. Jones has delineated the course of seventeenth century English scientific thought.[20] The English phase of the controversy largely concerns what we would term science, not the respective claims of ancient and modern literature, and the questions raised in the course of the controversy are sufficiently general that nearly every seventeenth- and eighteenth-century charge or claim of a scientific nature may be viewed as a contribution to the larger discussion. The lines of dispute are clearly drawn; the particular points at

20. I am indebted to such standard general studies of science in the period as A. Wolf, *A History of Science, Technology, and Philosophy in the 16th & 17th Centuries*, 2nd ed., rev. D. McKie (London, 1950); Wolf, *A History of Science, Technology, and Philosophy in the Eighteenth Century*, 2nd ed., rev. D. McKie (London, 1952); Herbert Butterfield, *The Origins of Modern Science: 1300–1800*, rev. ed. (New York, 1962); Alfred North Whitehead, *Science and the Modern World* (New York, 1925); Edwin Arthur Burtt, *The Metaphysical Foundations of Modern Physical Science*, rev. ed. (Garden City, 1954); A. R. Hall, *The Scientific Revolution, 1500–1800: The Formation of the Modern Scientific Attitude* (Boston, 1956), but the most useful treatment of the ideological tradition is to be found in Jones's *Ancients and Moderns*, in his "The Background of *The Battle of the Books*," *Washington University Studies*, 7, No. 2 (April 1920), 99–162, and in his "The Background of the Attack on Science in the Age of Pope," *Pope and His Contemporaries: Essays Presented to George Sherburn*, ed. James L. Clifford and Louis A. Landa (Oxford, 1949), pp. 96–113. An important recent study is Robert E. Schofield, *Mechanism and Materialism: British Natural Philosophy in An Age of Reason* (Princeton, 1970).

issue are simplified, and the extension of scientific or quasi-scientific ideas into other areas is invited and encouraged. Thus, even though the accomplishments of the experimental philosophers had won the day at the time of Swift's entering the lists, a writer such as Johnson could still announce himself as a Modern. The specific battle had been completed, the results decided, but the greater strife, though thrown into bold relief by the revolutions in science, is in a sense eternal, and the grounds of dispute cannot be limited by chronological boundaries since such questions as the relation between freedom and authority, deeds and words, induction and deduction, individual genius and collective accomplishments—all of which the combatants raised—defy final solution.

It is, however, misleading to exaggerate the eternal dimension of the conflict. Specific, pointed issues were raised. In elevating modern discoveries and inventions, pointing to the vulgar errors of classical and medieval science, advocating educational and stylistic reform, and seeking freedom from the shackles of "names" and "authority," the English "philosophers" offered a specific program of scientific study. Hinging on a methodology of observation and experiment, wary of systems and hypotheses, seeking utilitarian applications as well as theoretical victories, trusting in cooperative, collective effort, the Baconian program was largely adopted, though the basic inductive methodology would be employed in far more sophisticated fashion by Newton. Their goals and procedures were defended against quite real opposition, and the claims that science would enhance religion, provide new material for literature, aid in the amelioration of social ills, add further luster to English fame, and generally improve the quality of the national life, were made in high seriousness. The defense of science, indeed the very practice of science, was immediately related to this series of principles, claims, proposals, and presuppositions, which constituted the main material for the polemicist and ideological spokesman.

Though the ideological tradition is principally associated with Bacon, it is of course hardly restricted to his work. William Gilbert championed free investigation, denounced vulgar errors, and praised the moderns. Harvey professed "to learn and teach anatomy not from books but from dissections, not from the tenets of Philosophers but from the fabric of Nature."[21] Boyle was a tireless defender of the new science and prolific commentator on its relation with religion. Glanvill, Petty, Wilkins, and Henry Power may be numbered in the list of important commentators. The Milton of *Of Education* and the *Prolusions* belongs to the tradition, as does Evelyn, who suggested the motto of the Royal Society and designed the famous frontispiece to Sprat's history. Dryden contrasted the vitality of modern discoveries and methodology with the stagnation of ancient authority in his poem on Walter Charleton, and Charleton, in his *Physiologia Epicuro-Gassendo-Charltoniana* (which Butler attacks in his character, "A Philosopher"), spares no pains in characterizing the opponents of the "Assertors of Philosophical Liberty." The Aristotelians, for example, are the "FEMAL Sect; because as women constantly retain their best affections for those who untied their Virgin Zone; so these will never be alienated from immoderately affecting those Authors who had the Maiden head of their minds."[22] To Sir Thomas Browne, a "peremptory adhesion unto Authority" is "the mortallest enemy unto Knowledge."[23] The commentators may differ at some points—Power and Charleton, for example, proclaim an admiration for Descartes which their contemporaries did not all share—but they agreed on the jettisoning of antiquated error, the need for free investigation, and the successes of modern science.

Newton's comments on hypotheses in the General Scholium which follows Book III in the second edition of the *Principia* constitute one of the most famous ideological thesis statements in the tradition, and though he implicitly relies on the corpuscular hypothesis in his *Opticks*, he begins the work with a thoroughly representative announcement: "My Design in this Book is not to explain the Properties of Light by Hypotheses, but to propose and prove them by Reason and Experiments . . . ." In

21. William Harvey, *Movement of the Heart and Blood in Animals*, trans. Kenneth J. Franklin (Oxford, 1957), p. 7.

22. *Physiologia Epicuro-Gassendo-Charltoniana* (London, 1654), p. 2.

23. *Pseudodoxia Epidemica, Works*, ed. Geoffrey Keynes, II (London, 1964), 40.

the preface to his *Micrographia* Hooke comments at length on modern, and specifically English, science: Experimental philosophy is *real* philosophy; Peripatetick Matter and Form are useless words; the scientist, in cooperative fashion, should study tangible reality, not will-of-the-wisps; method is all important; the moderns can number gunpowder, the compass, printing, and microscopes among their inventions, and Harvey, Galileo, Ent, Wren, and Boyle among their colleagues; Hooke's age is an "inquisitive" one; the Royal Society is anti-rationalist; the Royal Society seeks experiments of fruit as well as light; and, finally, Hooke hopes that his work will contribute to a universal natural history. The preface is a veritable showcase of Baconian clichés and points up the tendency to state and restate the basic components of the contemporary scientific framework. The pattern repeats, of course, in the most famous post-Baconian ideological statement, Thomas Sprat's *History of the Royal Society of London, For the Improving of Natural Knowledge.*

Sprat's argument has been discussed often: in his "Advertisement to the Reader" he admits that he finds himself in the position of a defender, that the Society's detractors "make it necessary for [him] to write of it, not altogether in the way of a plain History, but sometimes of an Apology." The defense is an elaborate one. Hooke's principles are all restated and discussed in detail. The fears of the universities and the divines are allayed; the "Wits and Railleurs" are hopefully quieted; the universal advantages of experimental knowledge are demonstrated. Sprat's discussion of the relation between science and religion will be mentioned in a later chapter. At this point we must note his delineation of the Society's skeptical temper. He indicates that the Society's ideal lies between the extremes of dogmatism on the one hand and total skepticism on the other. What Johnson would term "incredulity" characterizes their tentative cast of mind. They approach their work with circumspection and modesty, examine rigorously, and withhold conclusions until the time is ripe and all the evidence is available.

In the eighteenth century the important defenders and popularizers of English science would be Dutch and French. In 1734 the fourth edition of Sprat's history appeared, as did the first volume of J. T. Desaguliers' *Course of Experimental Philosophy*, a popularization which we may use to indicate the two most important alterations in the ideological tradition between Sprat's time and Johnson's: the final toppling of Cartesian influence and the near deification of Newton. As Desaguliers notes in his Dedication to the Prince of Wales, the business of science is still "to contemplate the Works of God, to discover Causes from their Effects, and make Art and Nature subservient to the Necessities of Life . . . ." In his Preface he states that "All the Knowledge we have of Nature depends upon Facts; for without Observations and Experiments, our natural Philosophy would only be a Science of Terms and an unintelligible Jargon." Descartes had overthrown Aristotelian physics, but had unfortunately spawned "a new Set of Philosophers . . . whose lazy Disposition easily fell in with a Philosophy, that required no Mathematicks to understand it . . . ." However, a far worthier successor was soon to appear:

It is to Sir *Isaac Newton*'s Application of Geometry to Philosophy, that we owe the routing of this Army of *Goths* and *Vandals* in the philosophical World; which he has enrich'd with more and greater Discoveries, than all the Philosophers that went before him: And has laid such Foundations for future Acquisitions; that even after his Death, his Works still promote natural Knowledge.[24]

A goodly number of Goths and Vandals are still discernible in 1734, but the grand example and influence of Newton is undeniable. Three years after Desaguliers' work appeared, another writer would arrive in London whose praise of Newton and English science would be constant, and who would make a unique, though quiet contribution to the tradition of scientific ideology.

24. *A Course of Experimental Philosophy*, 1 (London, 1734), sigs. A2r, A7r, A7v.

CHAPTER II

Johnson in
an Age of Science

Early in 1775, in the *Monthly Review*, Ralph Griffiths noted the increased British interest in Scotland, and suggested that "tours to the Highlands, and voyages to the isles, will possibly become the fashionable *routes* of our virtuosi, and those who travel for mere amusement. Mr. Pennant has led the way; Dr. Johnson has followed . . . ."[1] The *Journey to the Western Islands of Scotland*, a work of considerable importance for the student seeking to delineate Johnson's intellectual temper, will be considered in the following chapter, but the suggestion that Johnson might justly be termed a virtuoso may be approved now. He does in fact share the scientific and antiquarian interests of many of his predecessors and contemporaries, and to the extent that such interest, coupled with experimentation, constitutes virtuosity, the reviewer's appellation is justified.

1. John Ker Spittal, ed., *Contemporary Criticisms of Dr. Samuel Johnson* (London, 1923), p. 161. For the identification of Griffiths as reviewer, see Benjamin Christie Nangle, *The Monthly Review: First Series, 1749–1789* (Oxford, 1934), p. 135.

It is unfortunate that Beauclerk did not provide us an extended memoir concerning his great friend, for he, unlike Boswell, shared Johnson's scientific interests. Boswell could not help noting Johnson's scientific avocations, particularly his enjoyment of amateur chemical experimentation, but though he often drew Johnson into discussions of his own favorite topics such as the law or the supernatural, he seldom encouraged conversation about natural philosophy. Further, his remarks on Johnson's scientific activities are not informed by a rigorous examination of Johnson's published statements on such activities. As a commentator he is here totally unreliable. By his own admission, he could not discern the hand of Johnson in James's *Medicinal Dictionary*,[2] where the most cursory examination would have at least uncovered the revised biography of Boerhaave. Because of his own lack of interest in science, many comments of Johnson have surely been lost. He does, however, call attention to some pertinent references in the *Diaries*,[3] where Johnson alludes to his philosophical pursuits. Boswell comments that Johnson "sometimes employed himself in chymistry, sometimes in watering and pruning a vine, and sometimes in small experiments, at which those who may smile, should recollect that there are moments which admit of being soothed only by trifles" (*Life*, III, 398).

The "small experiments" include Johnson's well-known empirical diversions: the weighing of dried leaves, the paring of nails and shaving of hair to judge rapidity of growth. He also amused himself with mathematical computations. In July, 1783, he writes to Sophia Thrale:

Nothing amuses more harmlessly than computation, and nothing is oftener applicable to real business or speculative enquiries. A thousand stories which the ignorant tell, and believe, die away at once, when the computist takes them in his gripe (*Letters*, III, no. 870, 54).

Mathematical amusements, in other words, are a means by which the man of affairs, the serious scientist, and enlightened

2. *Life*, III, 22, n. 6: "I have in vain endeavoured to find out what parts Johnson wrote for Dr. James. Perhaps medical men may."

3. *Samuel Johnson: Diaries, Prayers, and Annals*, ed. E. L. McAdam, Jr., with Donald and Mary Hyde (New Haven, 1958).

debunker can develop and enhance a shared device, and Johnson often records his penchant for making computations and taking measurements. Touring with the Thrales in 1774, for example, he notes lengths and breadths of the hall, gallery, library, and dining parlor at Llewenny; we learn that the Eagle Tower at Carnarvon Castle may be reached by ascending 169 ten-inch steps and that Woodstock park "contains 2500 Acres about four square miles. It has red deer" (*Diaries*, 184–85, 203, 221). Johnson—as each reader of the *Diaries* knows—continually attempts to keep track of his most minor financial dealings, and in the present case records costs and prices even though Thrale is covering the expenses.

This desire for exactness, which squares with his consistent demands for truthfulness and accuracy, constitutes what might be called a dominant theme in the *Journey*. In the *Diaries*, after observing pneumatic experiments at Salisbury, he includes calculations as to the area of a circle and the weight of a cubic foot of water (*Diaries*, 366), and the canon of his verse includes two Latin epigrams, one treating the proper proportion of a ship's load to the weight of water it can hold, the other asking and answering the question of the speed of sound (in feet/second).[4] The history of science, it need hardly be added, has not been dramatically affected by Johnson's mathematical and empirical propensities. If they are trivial in that they are not intended to lead to grand discoveries, they are still indicative of a particular cast of mind, and attempts to define Johnson's proclivities or ideological allegiances have not always considered these matters, and their far more important implications. It would be interesting to speculate on what might have been the early course of Johnson studies had he been imagined amid his chemical apparatus, rather than in his role of literary and moral dictator. Arthur Murphy first encountered him "all covered with soot like a chimney-sweeper, in a little room, with an intolerable heat and

---

4. *Samuel Johnson: Poems*, ed. E. L. McAdam, Jr., with George Milne (New Haven, 1964), p. 349. Cf. his "Geographia Metrica," ibid., pp. 350–51.

strange smell . . . making *aether*."[5] Whether in the garret over his chambers in the Inner Temple or in the laboratory which the Thrales provided for him at Streatham, Johnson the experimenter is in many ways far more appealing than the intellectually static man of prejudices, pontificating before the club, the counterfeit image inherited from the nineteenth century which Johnson's modern students have vigorously attempted to discard.

Johnson the traveler is particularly interested in "men and manners" but he seldom bypasses the opportunity to observe scientific collections or the technological extensions of science. After their visit, Mrs. Thrale was impressed by the museum of antiquities and curiosities of Richard Greene, the Lichfield apothecary,[6] and while in France Johnson visited the King's museum, which included insects and stuffed birds, the menagerie at Versailles, with which he was particularly taken, and the zoological collection and menagerie at Chantilly (*Diaries*, 231–32, 242, 255). His interests in practical applications of "philosophy," methods of manufacture, and mechanical inventions are by now proverbial. In his *Diaries* he records his trips, with the Thrales, to the silk mill at Derby (which would be repeated with Boswell in 1777), the brass, iron, and copper works at Holywell, Henry Clay's and Matthew Boulton's factories at Birmingham, the mirror factory at Paris, and the porcelain factory at Sèvres. In Nantwich they observed a process for obtaining salt through the evaporation of brine.[7]

"I have enlarged my notions," Johnson wrote, after seeing the metal works at Holywell, and his unabated curiosity continued despite the weight of years and failing health. The entry in his

5. See John J. Brown, "Samuel Johnson 'Making Aether'," *Modern Language Notes*, 59 (April 1944), 286. The reference is to Hester Lynch Piozzi, *Anecdotes of the Late Samuel Johnson, LL.D.*, ed. G. B. Hill in *Johnsonian Miscellanies* (Oxford, 1897), I, 306.

6. *Diaries*, pp. 163–64. Greene was a relative of Johnson, attended Lucy Porter, and was entrusted with the duty of placing an epitaph on the tombstone of Johnson's father, mother, and brother in St. Michael's Church.

7. Pertinent references include ibid., pp. 170–72, 186–87, 220, 243–44, 248, 277.

Aegri Ephemeris for October 20, 1784, is particularly gloomy, but on that day, less than two months before his death, he could still write to William Gerard Hamilton that he hoped "to find new topicks of merriment, or new incitements to curiosity" upon his return to London.[8] His strength and stamina precluded the possibility of illness or disease conquering curiosity, while his continual afflictions did incline him to the study of medicine, and, along with his proverbial gregariousness, brought him into the company of physicians, an important facet of his "scientific education"—to the degree that it can be defined—and one which he particularly enjoyed. In a famous letter to Edmund Hector, Johnson wrote, "My health has been from my twentieth year such as has seldom afforded me a single day of ease . . . ."[9] Equally telling is a diary entry for August 2, 1767: "I was extremely perturbed in the night but have had this day, 5–24 p.m. more ease than I expected. D[eo]. gr[atias]. Perhaps this may be such a sudden relief as I once had by a good night's rest in Fetterlane" (*Diaries*, 114). Since he resided in Fetter Lane in the years 1741–49, the infrequency of a single night's painless rest is clear. Modern physicians have retrospectively diagnosed the illnesses which beset Johnson as (among other things) agraphia, aphasia, asthma, chronic bronchitis, dropsy, emphysema, enlarged heart, gallstone, gout, scrofula, and slight hemiplegia. In addition he suffered from limited vision and poor hearing.[10] With this list in mind it comes as no surprise that Johnson was not only "a great dabbler in physick" (*Life*, III, 152), and prone to write his own prescriptions in technical language (in Montrose, for example,

8. *Letters*, III, no. 1024, 237.
9. Ibid., II, no. 772 (28 August ?), 474.
10. See, for example, Peter Pineo Chase, "The Ailments and Physicians of Dr. Johnson," *Yale Journal of Biology and Medicine*, 23 (April 1951), 370–79; Sir Humphry Rolleston, "Samuel Johnson's Medical Experiences," *Annals of Medical History*, NS 1 (Sept. 1929), 540–52; Russell Brain, "The Great Convulsionary" and "A Post-mortem on Dr. Johnson," in *Some Reflections on Genius and other Essays* (London, 1960), pp. 69–91, 92–100; Macdonald Critchley, "Dr. Samuel Johnson's Aphasia," *Medical History*, 6 (Jan. 1962), 27–44; Lawrence C. McHenry, Jr., "Dr. Johnson's Dropsy," *Journal of the American Medical Association*, 206 (9 Dec. 1968), 2507–9.

he was mistaken for a physician [*Tour*, p. 51]), but that he numbered among his acquaintances a large number of medical men.

As late as 1860 physicians and surgeons formed the largest group among the Fellows of the Royal Society. In 1698, 16 percent of the Fellows were medical figures, the number rising to 21 percent by 1740. The proportion is particularly striking in view of the fact that only about one-third of the Society's Fellows were noted for scientific concerns.[11] The conservatism of medicine is often cited, but it is undeniable that its practitioners were leading members of the eighteenth-century scientific world, and Johnson's acquaintance with a large number of them constitutes the most obvious evidence of an abiding interest in their art. A discussion of these friendships would make a lengthy study in itself and a cautious estimate of the number of Johnson's medical acquaintances would easily surpass three score.[12] I intend to mention only the closer ones in order to provide an indication of the number and variety of men with whom he came in contact.

Edmund Hector, one of Johnson's closest friends, was a Lichfield schoolmate and later a surgeon at Birmingham. Another important schoolmate was Robert James, author of the ponderous *Medicinal Dictionary* to which Johnson contributed, and a

11. Stimson, *Scientists and Amateurs*, p. 137; Sir Henry Lyons, *The Royal Society, 1660–1940: A History of its Administration under its Charters* (Cambridge, 1944), pp. 126, 341–42.

12. On Johnson and the medical profession, see: Chase, "The Ailments and Physicians of Dr. Johnson"; Rolleston, "Samuel Johnson's Medical Experiences"; H. E. Bloxsome, "Dr. Johnson and the Medical Profession," *Cornhill Magazine*, NS 58 (April 1925), 455–71; H. S. Carter, "Samuel Johnson and Some Eighteenth Century Doctors," *Glasgow Medical Journal*, 32 (July 1951), 218–27; F. N. Doubleday, "Some Medical Associations of Samuel Johnson," *Guy's Hospital Reports*, 101 (1952), 45–51; Robert Hutchison, "Dr Samuel Johnson and Medicine," *Edinburgh Medical Journal*, 97, Pt. 2 (August 1925), 389–406; Bertram M. H. Rogers, "The Medical Aspect of Boswell's 'Life of Johnson,' with Some Account of the Medical Men Mentioned in that Book," *Bristol Medico-Chirurgical Journal*, 29 (1911), 125–48; James P. Warbasse, "Doctors of Samuel Johnson and His Court," *Medical Library and Historical Journal*, 5 (1907), 65–87, 194–210, 260–72.

long-standing friend. Created M.D. in the University of Cambridge by royal mandate, James practiced in Sheffield, Lichfield, and Birmingham before settling in London. Goldsmith, another doctor and one very close to Johnson, comes next to mind, for he is reputed to have died from taking James's famous fever powder. Johnson admired James's abilities but, appropriately, had severe reservations about his compound medicines. The less complex methods of a man like Sydenham, who tended to trust natural processes far more than the concoctions of human hands, were closer to Johnson's beliefs and opinions. Nevertheless, the fever powder was the most popular medicine of the day and was used in one of George III's attacks of mental disease and in the last illness of George Washington. Samuel Swinfen, one of Johnson's godfathers, was a physician, as was Joseph Ford, Johnson's uncle and the father of the famous parson, so that Johnson's early contacts with physicians established a pattern which would continue throughout his life.

Dr. Richard Bathurst, a close and beloved friend, contributed to the *Adventurer* and was a member of the Ivy Lane Club. An efficient but unsuccessful physician, he later became an army doctor in an expedition against Havana. Francis Barber was once his father's slave in Jamaica. Thomas Lawrence, one of Johnson's personal physicians and dearest friends, was introduced by Bathurst; President of the Royal College of Physicians, Johnson praised him as "venerable for his knowledge, and more venerable for his virtue."[13] Robert Levet's name, far humbler than Lawrence's, is immediately associated with Johnson's elegy. Originally a waiter in a Paris coffee house, Levet impressed its medical patrons to the extent that they provided funds as well as advice and launched his modest career. In London he attended Hunter's

13. *Letters*, I, no. 367 (20 Dec. 1774 to Warren Hastings), 420. Howard D. Weinbrot has informed me that Johnson corrected individual passages of Dr. Lawrence's unpublished "De natura animali dissertatio," the holograph of which is preserved in the Harveian Library of the Royal College of Physicians. Some of Johnson's corrections (referring to pp. 23–32, 40–43 of the MS) are annexed to this volume; others (referring to pp. 1–6) are now at the Huntington Library. Weinbrot suspects that at least one more sheet of remarks (covering pp. 7–22) is yet to be found. He will publish in the near future the materials which have come to light.

lectures and practiced among the poor and hopelessly indigent. If he has been called a quack,[14] Johnson's opinion of his medical abilities was high (*Life*, I, 243).

In the imaginary college which Johnson and Boswell whimsically planned to establish at St. Andrews, Dr. Christopher Nugent was to teach medicine (*Tour*, p. 78). Nugent was Burke's father-in-law, and, along with Drs. George Fordyce and Richard Warren, was a member of the Literary Club. Warren enjoyed a very remunerative practice and numbered George III among his patients. He attended Johnson in his last illness, as did William Heberden, the first to describe angina pectoris and perhaps the outstanding internist of the time. Heberden was a member of the Essex Street Club, a prolific writer, and physician to William Cowper. Johnson was also attended at this time by Drs. Richard Brocklesby, William Cruikshank, and William Butter. Brocklesby, a physician to Burke and Wilkes, was also a member of the Essex Street Club. A man of kindness, generosity, and considerable wealth, he was a learned and lively conversationalist. Cruikshank, one of the most prominent surgeons in the second half of the century and an able anatomist, was William Hunter's partner. He also attended Johnson in 1773 for hydrocele[15] and was recommended by him as successor to Hunter's Professorship of Anatomy at the Royal Academy. Before he moved to London, Butter's practice was at Derby, where he entertained Johnson and Boswell and conducted them through the much-admired china factory.

Sir William Browne, the physician who disputed with Johnson the question of Oxford's superiority to Cambridge, is known to all readers of Boswell, as is John Coakley Lettsom, who was present at the famous meeting with Wilkes. In 1762 Johnson and Reynolds visited John Mudge, the Plymouth surgeon; Amyat, a London physician, was also encountered during this Devonshire

14. Chase, "The Ailments and Physicians of Dr. Johnson," p. 377.

15. With Donald Greene, I follow H. S. Carter, who follows Johnson himself in diagnosing hydrocele rather than sarcocele. See Greene, "Dr. Johnson's 'Late Conversion': A Reconsideration," in *Johnsonian Studies*, ed. Magdi Wahba (Cairo, 1962), p. 64, n. 8; Carter, "Samuel Johnson and Some Eighteenth Century Doctors," pp. 223–24.

jaunt. Percivall Pott, one of the most eminent of London surgeons, gave his name to the disease from which Pope suffered, and, with Cruikshank, attended Johnson for hydrocele. Johnson was also seen occasionally by Sir George Baker, Reynolds' physician, and Sir Lucas Pepys, the family physician to the Thrales. This list does not pretend to be complete. We have not mentioned James Grainger, the poet-physician whose practice in the Leeward Islands failed; Dr. John Boswell, Boswell's uncle, "a physician bred in the school of Boerhaave" (*Tour*, p. 385); Dick, Gillespie, Cullen, Hope, and Monro, whom Boswell, in Scotland, consulted during Johnson's final illness; William Barrowby, who wished he were a Jew in order to enjoy swine's flesh and the pleasure of sinning simultaneously; Dr. Samuel Musgrave, whose poem "The Project" Johnson sharply criticized; or the apothecaries Holder and Diamond, but the list is, I trust, sufficiently inclusive to indicate an extremely large number of dealings with members of the medical profession. Since Anna Williams and Elizabeth Desmoulins also required frequent medical aid, Johnson's contacts with physicians must have been nearly constant. Levet attained much of his medical knowledge through personal contacts and conversation; Johnson was surely able to do the same.

Among other men of science with whom Johnson associated, the closest and most famous acquaintance was Sir Joseph Banks. President of the Royal Society from 1778 to 1820, Banks was chiefly interested in natural history, particularly botany, horticulture, and agriculture. He was a member of the Literary Club and one of Johnson's pall bearers. Johnson speaks of him with admiration and at one time planned to join him on one of his expeditions.[16] Topham Beauclerk, as was indicated earlier, shared

16. For Johnson's admiration, see *Letters*, II, no. 587 (31 Oct. 1778 to Bennet Langton), 264; II, no. 593 (21 Nov. 1778 to Boswell), 272. On the expedition, see *Life*, II, 147–48. In 1771 the Admiralty desired an expedition to explore the high southern latitudes. Two ships, the *Resolution* and the *Adventure*, were to be commanded by Captain James Cook. Banks wanted to join them and was willing to go at his own expense, but the Admiralty finally refused. Banks's equipment and party would

Johnson's scientific curiosity. He studied floriculture and astronomy, had a taste for rare books and collected one of the most splendid private libraries in England, numbering over 30,000 volumes. Johson "was entertained with experiments in natural philosophy"[17] at his laboratory at Windsor; in the imaginary college at St. Andrews it was he who would teach science.

In 1760 Johnson met two famous scientists. Between May and December he met on three occasions with Roger Boscovich, the Serbian physicist and mathematician. At Mrs. Cholmondeley's they discussed Newton (Boscovich was one of the first foreign scientists to adopt the Newtonian theory of gravitation) and Johnson characteristically maintained "the superiority of Sir Isaac Newton over all foreign philosophers" (Life, II, 125). Boscovich was indebted to Johnson for a letter of introduction to Robert Chambers, which contributed much to his cordial reception at Oxford.[18] In May, 1760, as Maurice Quinlan has demonstrated,[19] Johnson met Benjamin Franklin. Both were members of a semi-religious benevolent association called the Associates of Dr. Bray. What they discussed—if, indeed, a conversation occurred—we do not know, but, as with Boscovich, Johnson may have enjoyed the stimulation of speaking with an important scientist. Quinlan speculates that they had possibly been in each

---

have required too extensive an alteration of the ship. Banks changed his plans, chartered his own vessel, and set out on an expedition to Iceland. See Lyons, The Royal Society, pp. 189–90. For an extended study of Banks, see Hector Charles Cameron, Sir Joseph Banks, K.B., P.R.S., The Autocrat of the Philosophers (London, 1952).

17. Life, I, 250. On Beauclerk, see Frederick M. Smith, "An Eighteenth-Century Gentleman: The Honorable Topham Beauclerk," Sewanee Review, 34 (April 1926), 205–219. A great-grandson of Charles II, Beauclerk was said to resemble the king in appearance as well as in morals; it is interesting to note that Johnson displayed a strong affection for Charles, the protector of the Restoration experimenters and charterer of the Royal Society. See Life, II, 341.

18. Josip Torbarina, "The Meeting of Bošković with Dr. Johnson," Studia romanica et anglica Zagrabiensia, Nos. 13–14 (1962), p. 5.

19. "Dr. Franklin Meets Dr. Johnson," New Light on Dr. Johnson, ed. Frederick W. Hilles (New Haven, 1959), pp. 107–120.

other's company before this meeting and that it is very likely that they saw each other on later occasions. We do know that Johnson was conversant with Franklin's electrical experiments. They had received considerable publicity in the *Gentleman's Magazine*; four years before their meeting Johnson had written in the *Literary Magazine*:

> there is no reason to doubt that the time is approaching when the *Americans* shall in their turn have some influence on the affairs of mankind, for literature apparently gains ground among them. A library is established in *Carolina*; and some great electrical discoveries were made at *Philadelphia* . . . .[20]

We may add to Quinlan's speculations concerning the occurence of several meetings the suggestion that these "great electrical discoveries" may well have been a topic of discussion.

Throughout his life Johnson associated with men intimately or remotely connected with the new science. In Glasgow he dined with John Anderson, Professor of Natural Philosophy (*Tour*, p. 365). He was a friend of Thomas Birch, a secretary of the Royal Society, whose history of the Society he reviewed in the *Literary Magazine*. We find him dining with John Henderson, "celebrated for his wonderful acquirements in Alchymy, Judicial Astrology, and other abstruse and curious learning" (*Life*, IV, 298), meeting Sir Alexander Gordon, Professor of Physick at King's College, Aberdeen,[21] and riding in a coach, enjoying "much instructive conversation," with John Hope, Professor of Botany at Edinburgh (*Tour*, p. 393). We see him attending chemical experiments in Wiltshire (*Life*, IV, 237–38), writing to Mrs. Thrale of his having seen a "great Burning Glass,"[22] taking part in the establishment of an early provincial infirmary,[23] interested in the construction of a roller-spinning

20. Review of Evans' map and account of the American middle colonies, *Literary Magazine*, 1, vi, 293.

21. *Letters*, I, nos. 321, 322, 326 (25 Aug., 28 Aug., 15–21 Sept., 1773, to Mrs. Thrale), 346, 354.

22. Ibid., III, no. 858 (30 June 1783), 42.

23. See E. L. McAdam, Jr., and A. T. Hazen, "Dr. Johnson and the Hereford Infirmary," *Huntington Library Quarterly*, 3 (April 1940), 359–67.

machine,[24] and even subscribing to a balloon-ascent project.[25]

One of John Brown's main points concerning Johnson and eighteenth-century science is his interest in technology. Johnson's ability to grasp quickly the principles underlying mechanical contrivances and processes of manufacture has often been noted, as has his approval of technological advances in his period. He was a member of the Society of Arts, the aim of which was to encourage practical applications of theoretical science. The Society promoted arts, manufactures, commerce, and agriculture; H.M.S. *Bounty*, for example, began its expedition under the Society's encouragement. Recently Professor John Abbott has shown that Johnson's connections with the Society were relatively extensive. There are some twenty-five specific references to him in the Society's records; he served on at least five Society committees and proposed four men for membership.[26]

When Brown notes Johnson's technological interests he is on perfectly safe ground. However, his estimate of Johnson's scientific knowledge as a whole is open to question. Brown concluded that Johnson's knowledge of medicine probably compared favorably with that of a contemporary provincial physician, that he knew more chemistry than the average well-informed layman of the period, that his knowledge of natural history was sparse, his acquaintance with biology, mathematics, physics, and economics not exceeding what we would expect from an intelligent, curious, professional writer.[27] I would argue that Brown's estimate can neither be confirmed nor denied. Johnson seldom makes technical assertions which would enable us to assess his strengths or limitations. Since an appraisal of his scientific com-

24. See John J. Brown, "Samuel Johnson and the First Roller-Spinning Machine," *Modern Language Review*, 41 (Jan. 1946), 16–23; and J. de L. Mann, "Dr. Johnson's Connection with Mechanical Spinning," ibid. (Oct. 1946), 410–11.

25. *Letters*, III, nos. 929.1 (31 Jan. 1784 to Hester Maria Thrale), 128; 929.2 (3 Feb. 1784 to William Bowles), 129–30; 997 (21 Aug. 1784 to Richard Brocklesby), 204–5.

26. "Dr. Johnson and the Society," *Journal of the Royal Society of Arts*, 115 (April, May 1967), 395–400, 486–91.

27. John J. Brown, "Samuel Johnson and Eighteenth-Century Science," Diss. Yale 1943, pp. 270–72.

petence is almost entirely based on inference, there is no foundation for a certain assessment, and the possibility is always great
that the uncovering of new material will radically alter assess-
ments that are admittedly tentative. A case in point is Johnson's
interest in electricity. Professor Wimsatt has noted the thoroughgoing competence of the *Dictionary* definition of "electricity," which indicates that Johnson had kept up to date with
relevant developments in the field. [28] Quinlan has established the
facts concerning the meeting with Franklin; Donald Greene has
suggested that Johnson may have reviewed Lovett's *The Subtil
Medium Proved* and Hoadly and Wilson's *Observations of a
Series of Electrical Experiments* for the *Literary Magazine*.[29] We
know definitely that he recommended the therapeutic application of electricity for the ailing Dr. Lawrence.[30] Zachariah Williams and Stephen Gray had been friends in the Charterhouse, as
Johnson notes in the pamphlet written on Williams' behalf;
Anna Williams, acting as assistant to Gray, claimed to be the first
to observe the emission of an electrical spark from a human
body, and Johnson is almost wholly responsible for "her" poem,
"On the Death of Stephen Grey, F.R.S., The Author of the Present Doctrine of Electricity." Here, in other words, is ample evidence of more than a passing interest in electricity. As our
knowledge of Johnson's life and writings increases, such matters
are continually brought to light, but since Johnson's tendency is
to place scientific material in broad human contexts, to avoid the
technical and specific in favor of the universal, we are not ever
likely to approach the point at which we can clearly define the
extent of his learning. The course of his "scientific education"
can be traced in skeletal form, but its exact details generally remain elusive, and when they are not elusive they are not always
easy to evaluate. Besides such triumphs as the definition of "elec-

28. "Johnson on Electricity," *Review of English Studies*, 23 (July 1947),
257–60.
29. "Johnson's Contributions to the *Literary Magazine*," ibid., NS 7
(Oct. 1956), 386–88.
30. *Letters*, ɪɪ, no. 802 (26 Aug. 1782 to Elizabeth Lawrence), 504: "I
should not despair of helping the swelled hand by electricity, if it were
frequently and diligently applied."

tricity" there are blunders such as Johnson's comments on microscopy to George III or his famous claim that swallows "conglobulate together, by flying round and round, and then all in a heap throw themselves under water, and lye in the bed of a river" (*Life*, II, 55). However, Linnaeus also claimed that swallows do not migrate but rather spend the winter under water in ponds.[31] But Linnaeus refused ever to change his opinion, while Johnson, when he was told, five years later, that there is indeed migration of swallows remained silent.[32] His teacher on this occasion, however, was Goldsmith, whose knowledge of natural philosophy has never been judged worthy of emulation.

During the course of his career Johnson was engaged in three projects which involved the reading or writing of scientific works and which served to enhance his extensive self-education. The first was the "collaboration" with Dr. Robert James on his *Medicinal Dictionary* of 1743–45. For this work Johnson wrote the Proposals, Dedication to Dr. Richard Mead, and (at most, I think) a dozen articles, chiefly biographies of physicians.[33] To write his medical biographies he used such conventional sources as LeClerc's *Histoire de la Médecine* and Fontenelle's *Éloges*.[34] How much he actually learned from the project is impossible to estimate, but his remark that he acquired his "knowledge of physick" from James (*Life*, III, 22) indicates that this experience along with other contacts with James were especially valuable. At approximately the same time as the work for James, Johnson was engaged in the cataloguing of the massive Harleian library, the results of which appeared in 1743–44. Some 5,000 of the vol-

31. Charles E. Raven, *Natural Religion and Christian Theology*, I (Cambridge, 1953), 156.

32. *Life*, II, 248. See Robert Donald Spector, "Dr. Johnson's Swallows," *Notes and Queries*, 196 (22 Dec. 1951), 564–65.

33. See Allen T. Hazen, "Samuel Johnson and Dr. Robert James," *Bulletin of the Institute of the History of Medicine*, 4 (June 1936), 455–65; Hazen, "Johnson's Life of Frederic Ruysch," ibid., 7 (March 1939), 324–34; and Lawrence C. McHenry, Jr., "Dr. Samuel Johnson's Medical Biographies," *Journal of the History of Medicine and Allied Sciences*, 14 (1959), 298–310.

34. John L. Abbott, "Dr. Johnson's Translations from the French," Diss. Michigan State, 1963, p. 119.

umes dealt in one way or another with science.[35] Again, it is difficult to judge how much Johnson was able to glean in the course
of the cataloguing, but it would surely be obtuse to assume that
a man of Johnson's abilities, possessing voracious curiosity, and
experienced as well as capable in reading bits and snatches, would
have seen the work as merely mechanical drudgery and learned
little in the process.

The third and by far the most important of the projects which
were in some way connected with science was, of course, the
compiling of the *Dictionary*. In preparing his work, as Wimsatt
has demonstrated,[36] Johnson consulted such scientific writers as
Arbuthnot (for whom he held great admiration), Bacon, Boerhaave (whose life he had written in 1739 and extensively revised
for the *Medicinal Dictionary*), Boyle, Musschenbroek, Newton,
Wilkins, and the much-maligned John Woodward. Among the
psysico-theological writers, he used Bentley's Boyle lectures,
Blackmore's *Creation*, Burnet's *Theory of the Earth*, Cheyne's
Philosophical Principles of Religion, Natural and Revealed, Der-
ham's *Physico-Theology* and later *Astro-Theology*, Nehemiah
Grew's *Cosmologia Sacra*, and Ray's famous *Wisdom of God
Manifested in the Works of the Creation*. He used Sir Thomas
Browne (whose life he was soon to write), Glanvill, Locke, and
Chambers' very important *Cyclopaedia*. In short, he surveyed a
broad spectrum of scientific writing and the knowledge gained
must have been great indeed. Wimsatt argues that, even if Johnson was only casually attentive in reading for the *Dictionary*, this
type of reading is often productive of side discoveries and in-
spirations.[37] We should remember also that Johnson forgot very
little. A man who can quote from memory poetry that he dis-
liked can make casual or even inattentive reading a very important pursuit.

35. Brown, "Samuel Johnson and Eighteenth-Century Science," p. 57;
W. K. Wimsatt, Jr., *Philosophic Words: A Study of Style and Meaning
in the "Rambler" and "Dictionary" of Samuel Johnson* (New Haven,
1948), p. 51.

36. See Appendix B, "Some Philosophic Sources of Johnson's Dictionary," ibid., pp. 146–60.

37. "Johnson's Dictionary," in *New Light on Dr. Johnson*, pp. 83–84.

The most intriguing index to Johnson's self-education which has survived is the sale catalogue of his personal library. The catalogue sometimes raises more problems than it solves. Many lots are not described; individual titles are often listed with little regard for accuracy; works that we can be sure Johnson knew, and many that he knew well, are not to be found. We may also assume, with S. C. Roberts,[38] that many books with personal associations, books of special importance to Johnson, were bequeathed verbally, as well as in writing, to friends prior to his death. Moreover, we cannot be certain that Johnson read all the books he possessed, and whether his reading was intense or dilatory can seldom be ascertained. We can assume however that he at least inspected the great majority of them and we can be certain that a casual inspection on his part can be equated with a prolonged study by an average reader. Since we know that he enjoyed access to the libraries of his friends as well as to public resources, that a vast number of scientific works passed through his hands in the cataloguing of the Harleian collection and the preparation of the *Dictionary*, that he was forced to consult various sources for his contributions to the *Medicinal Dictionary*, and that he wrote book notices and reviews for the *Gentleman's* and the *Literary Magazine*, the sale catalogue would only call into question his scientific learning if scientific material was conspicuously absent from the lists. We know that he was exposed to scientific writings for extended periods and can demonstrate the fact without even looking at a catalogue of his personal library, but the catalogue fully justifies the assumption that his scientific reading was extremely broad and that his ability to comment on science is thoroughly supported by first-hand acquaintance with primary material. Moreover, the gaps in the catalogue are sufficiently obvious to justify the inference that we are dealing with the proverbial iceberg's tip. The number of scientific works in Johnson's possession which the catalogue lists provides only a bare indication of the wealth of resources on which he could base his generalized commentary.

An actual count of the scientific and medical works in John-

38. "Johnson's Books," *The London Mercury*, 16 (Oct. 1927), 618.

son's library is virtually impossible, since many writers (Locke, Barrow, Descartes, Aristotle, Samuel Clarke, for example) could be placed in numerous categories and the library included text-books, quasi-scientific compendia, popularizations, reference works, and philosophic and literary works closely related to nat-ural philosophy. With the list of books that Johnson took to Ox-ford[39] we are on surer ground. The proportion of "scientific" works (Lucretius, Blackmore's *Creation*, Garth's *Dispensary*, John Harris' *Description . . . of the . . . Globes*) is very slight. But Roberts' estimate that of the 650 lots in the sale catalogue (representing 3,000–5,000 volumes), nearly 60 relate to mathe-matics, science, and medicine,[40] is extremely conservative and could easily be doubled.

The largest group of books is, as would be expected, medical. It includes the works of classical physicians such as Galen, Are-taeus (whose life Johnson wrote), Hippocrates, and Celsus, who Johnson particularly admired. Among the moderns there are Fabricius, Fallopius, and Vesalius; among contemporaries, Mead, Haller (perhaps Boerhaave's most famous student), Robert James of course, and Percivall Pott, one of Johnson's personal physicians. LeClerc's history of medicine, used in preparation of some of the medical biographies, is there, as are Salmon's *Prac-tice of Curing*, Hill's *Materia Medica*, Mudge's *On a Catarrhous Cough*, Cheselden's *Anatomy*, and Cheyne's *English Malady*, the last a particular favorite. The Hunters, Thomas Sydenham, William and Gideon Harvey are not present. Johnson had used Gideon Harvey for the *Dictionary* and written a biography of Sydenham, so that the shortcomings of the catalogue as an in-clusive index should be immediately manifest.

The list of chemical writers includes Boyle (whose work Johnson knew and continually praised), Boerhaave, Paracelsus, J. J. Becher, G. E. Stahl, and van Helmont, so that Johnson sure-ly knew the attempts of the iatro-chemists to put chemistry at the service of medicine, the propensities of the spagyrists, the grounds for dispute over the heuristically useful phlogiston the-

39. *Johnsonian Gleanings,* ed. Aleyn Lyell Reade, v (privately printed, 1909–1952), 213–29.

40. "Johnson's Books," p. 616.

ory, and some of the developments in pneumatic chemistry. Priestley's name, it should be noted, is absent. In physics, mechanics, and astronomy, the works of Archimedes and Ptolemy are present, as is Gilbert's *De Magnete*. Newton is there of course, as are Musschenbroek and Gassendi, so that Johnson would have a first-hand awareness of the modern claims of Epicurean atomism, though Hobbes's and Walter Charleton's atomistic works are not represented. In natural history, Johnson's library includes the works of Aristotle and Pliny, and among the moderns those of Cesalpino, Conrad Gesner, Nehemiah Grew, and Stephen Hales, a man Johnson justly admired and some of whose work he reviewed. The greatest hiatus here, and indeed throughout the catalogue, is the work of Bacon, which Johnson knew, praised, continually cited, and even considered editing.

Among the scientific philosophers and men whose work is, in various ways, closely connected with science or its methodology, we find Descartes, Bayle, Locke, Clarke (Johnson knew of and took sides on the Leibniz-Clarke controversy), Barrow, Hartley, and Locke's follower, Isaac Watts. Derham and Pascal, whose works Johnson knew and recommended, are not listed. Nor are Berkeley and Henry More, both of whom Johnson cited in the *Dictionary*. Among the geologists, Johnson possessed the writings of Woodward and Burnet. Both were used in the *Dictionary*; Woodward, it will be seen, was defended by Johnson against the charges of the Scriblerians. Maupertuis' *Figure of the Earth* is also represented, so that the oblate spheroid–prolate spheroid controversy and the extensive work it generated would not be foreign to him. Cocker's *Arithmetic*, a book presented as a gift on the Hebridean jaunt,[41] is not present, but the geometric and trigonometric books of William Payne, for which Johnson wrote Dedications, are there, as are, again, Barrow's and Newton's works, and 'sGravesande's *Elementa Mathematica*.

Miscellaneous works deserving of mention include Arbuthnot's tables of ancient coins, Chambers' *Cyclopaedia*, 16 volumes of Leibniz's journal, the *Acta Eruditorum*, 7 volumes of the *En-*

41. *Tour*, p. 104. See Stephen O. Mitchell, "Johnson and Cocker's *Arithmetic*," *Papers of the Bibliographical Society of America*, 56 (Jan.-March 1962), 107–9.

cyclopédie,[42] Birch's history of the Royal Society, Browne's *Pseudodoxia Epidemica*, the works of Roger Bacon, Baptista Porta, and Antonio Neri's *De Arte Vitraria*, one of several possible sources for the magnificent discussion of glass in the ninth *Rambler*.[43] Blackmore and Glanvill, both of whom Johnson knew and used in the *Dictionary*—Blackmore's *Creation* was not only taken to Oxford, but was to be singled out for special praise in the *Lives of the Poets*—are absent, as is, surprisingly, Sprat's history. Of particular interest is the relative lack of material on electricity.

At this point it should be clear that Johnson's personal friendships, social contacts, professional literary undertakings, and private self-education all contributed to the acquisition of a body of scientific learning. His "philosophic" dabbling and scientific reading are thoroughly characteristic of the age to which he gave his name, but the form in which his learning found expression is unique, not because his perception of the interrelations of science and other human concerns is original, but because it is articulated in works bearing the stamp of genius. Professor Wimsatt has justly observed that Johnson's discussions of science are seldom specific; they are general rather than technical, but their "generality . . . readily allies itself with the broader and more easily available features of seventeenth-century and eighteenth-century science."[44] To his role as commentator we must now turn.

In 1714 the British government offered a prize of £20,000 for any successful method of determining longitude at sea within 0.5°, with smaller prizes for less accurate methods. By 1828 the Board of Longitude had paid more than £100,000 in assistance and rewards to inventors.[45] The number of individuals who de-

42. See L. F. Powell, "Johnson and the *Encyclopédie*," *Review of English Studies*, 2 (July 1926), 335–37.

43. Wimsatt, *Philosophic Words*, pp. 76–77.

44. Ibid., p. 79.

45. See Wolf, *A History of Science . . . in the Eighteenth Century*, pp. 153–60; W. E. May, "Longitude by Variation," *Mariner's Mirror*, 45 (Nov. 1959), 339–41; E. G. R. Taylor, "A Reward for the Longitude," ibid. (Feb. 1959), 59–66; and Edmond Guyot, *Histoire de la Détermination des Longitudes* (La Chaux-De-Fonds, 1955).

voted time and effort to the problem included Hooke, Newton, Halley, and—most important for our purposes—Dr. Zachariah Williams. Williams had given up his medical practice in Wales to attempt a solution based on magnetic variation. Such a method had been advanced by Gilbert in the days of Elizabeth and was tried in the eighteenth century by, among others, John Philip Barretier, as Johnson notes in his biographical sketch of that famous prodigy.[46] Williams doubted the possibility of constructing more accurate apparatus for the measurement of time and rejected that approach, the one that was to win the prize for John Harrison of Yorkshire (1693–1776). Harrison's fourth marine timepiece, the result of continual refinement, was extremely efficient. When we turn to the work of Pierre LeRoy, whom Johnson met during the course of his 1775 travels in France with the Thrales, we see an invention embodying most of the essential features of the modern chronometer.

In 1755, *An Account of an Attempt to Ascertain the Longitude at Sea . . .* , a pamphlet written by Johnson on Williams' behalf, appeared. John J. Brown considered it Johnson's major scientific work.[47] Such a claim, I would argue, involves a basic misunderstanding of the nature of Johnson's scientific commentary. There is no basis for assuming that Williams' ideas are shared by Johnson. In view of the *Life of Barretier* we can be certain that Johnson knew of at least one unsuccessful attempt based on magnetic variation, a method which had been judged unworkable as early as the 1630s. Moreover, in his diary of the French tour, Johnson's notice of LeRoy is cordial;[48] there is, to be sure, no acrimony or envy. Far from being certain, the "Williams" of the pamphlet includes careful qualifications of his claims:

By the hope of sudden riches many understandings were set on work very little proportioned to their strength, among whom

46. 1825 *Works*, vi, 389.

47. "Samuel Johnson and Eighteenth-Century Science," p. 77.

48. *Diaries*, p. 234: "Then we went to Julien [Johnson's slip of the pen; Julien was Pierre's father] Le Roy the King's Watchmaker, a man of character in his business who shewed a small clock made to find the longitude. A decent man."

whether mine shall be numbered, must be left to the candour of posterity . . . .

In this state of dereliction and depression, I have bequeathed to posterity the following table; which, *if time shall verify my conjectures*, will show that the variation was once known . . . (1825 *Works*, v, 295, 301 [my italics]).

Johnson's own knowledge of navigational problems and possible solutions cannot be impugned on the basis of this work; the pamphlet does, however, represent a further instance of Johnson's charity and humanity. Considering the fact that he presented a copy of the pamphlet to the Bodleian and that it contained an Italian translation by Baretti, I think it safest to say that Johnson was merely doing all in his power to give Williams' ideas a fair and wide hearing. Johnson's constant desire is to show the ways in which men involved in the studies peculiar to different disciplines share common ground, or the ways in which a particular experience of a single man or the principles of a particular discipline are relevant for all men. The Williams pamphlet is a memorable commentary on human aspirations, but it is Johnson the moralist, not the would-be scientist, who reveals himself in such passages as the following:

Thus I proceeded with incessant diligence; and, perhaps, in the zeal of inquiry, did not sufficiently reflect on the silent encroachments of time, or remember, that no man is in more danger of doing little, than he who flatters himself with abilities to do all. When I was forced out of my retirement, I came loaded with the infirmities of age, to struggle with the difficulties of a narrow fortune; cut off by the blindness of my daughter from the only assistance which I ever had; deprived by time of my patron and friends; a kind of stranger in a new world, where curiosity is now diverted to other objects, and where, having no means of ingratiating my labours, I stand the single votary of an obsolete science, the scoff of puny pupils of puny philosophers (1825 *Works*, v, 301).

In December, 1759, Johnson wrote three letters which appeared in the *Gazetteer* concerning the proposed methods for constructing Blackfriars Bridge. John Gwynn, the architect who designed Magdalen Bridge and other structures at Oxford, induced Johnson to enter the controversy, the chief point of contention of which was whether the bridge should have semicircular or elliptical arches. Johnson, arguing on behalf of Gwynn, proposed semicircular arches; the elliptical arches were chosen and, once again, we find Johnson defending a losing cause for a friend. Since he considered architecture one of the "kindred arts" of mathematics[49] these letters may be considered attempted contributions to applied science articulated in a technical framework. Johnson can never resist the tendency to generalize (letter II, e.g., begins: "In questions of general concern, there is no law of government, or rule of decency, that forbids open examination and publick discussion.") but here he is writing science, not writing about it. I mention the letters to emphasize by contrast the fact that the great majority of his writings on science are commentaries which reflect attitudes, not attempts to describe formal systems, experiments, or discoveries. The letters, and to an extent the longitude pamphlet, illustrate the important point that Johnson's relation to the history of science cannot be defined in terms of any technical contribution. The nature of his statements suggests that our interest should lie in the ways in which he viewed science and the manner in which it affected his mind and art, not in his own scientific activities. He is a formidable commentator, but not a scientist.

Johnson's reading in the literature of science, his acquaintance with physicians, lay and professional "philosophers," and his private experimental dabbling, found expression in numerous ways; his reflections and pronouncements took many forms. His high opinion of and preference for biography is a Boswellian commonplace; it comes as no surprise that many of his statements on scientists, their studies, methodology, and extra-scientific presuppositions, occur in biographical writings. Early in his career Johnson wrote biographical sketches of three medical figures for the *Gentleman's Magazine*: Hermann Boerhaave (1739), the Dutch physician and scientist; Lewis Morin (1741), a French physician and botanist who has recently been suggested as a prototype of the astronomer in *Rasselas*;[50] and Thomas Sydenham

49. 1825 *Works*, v (*An Account of the Harleian Library*), 187.
50. F. V. Bernard, "The Hermit of Paris and the Astronomer in *Rasselas*," *Journal of English and Germanic Philology*, 67 (April 1968), 272–? `

(1742), the renowned English physician and close friend of Locke.[51] The *Life of Boerhaave* (the most ambitious and, I think, the most important of these biographies) was revised and expanded for Dr. James's *Medicinal Dictionary* to include discussions of several of Boerhaave's works. A slightly abbreviated version of the *Medicinal Dictionary* biography was later printed in the *Universal Magazine* (see Appendix A). The *Life of Sydenham*, besides appearing in the *Gentleman's*, served as a preface to Dr. John Swan's 1742 edition of Sydenham's works.

In the massive *Medicinal Dictionary*, which has been described as "the largest, most exhaustive and most learned medical dictionary written in English in pre-scientific days,"[52] the following biographical entries have been attributed to Johnson: "Actuarius," "Aegineta," "Aesculapius," "Aetius," "Alexander," "Aretaeus," "Archagathus," "Asclepiades," "Oribasius," "Ruysch," and "Tournefort."[53] The historical and biographical section of the article "Botany" has also been suggested as Johnson's, but I consider part of this attribution doubtful (see Appendix B). In 1756 Johnson wrote what Boswell considered one of his "best biographical performances" (*Life*, i, 308), a life of Sir Thomas Browne, prefixed to the second edition of Browne's *Christian Morals*. The *Lives of the Poets*, though its main interest is literary rather than scientific, includes sketches of the poet-physicians Cowley, Garth, Blackmore, and Akenside, as well as

51. On this biography, see Lawrence C. McHenry, "Samuel Johnson's 'The Life of Dr. Sydenham'," *Medical History*, 8 (April 1964), 181–87. Dr. McHenry corrects some of the errors in Johnson's sketch.

52. Charles W. Burr, "Dr. James and his Medical Dictionary," *Annals of Medical History*, NS 1 (March 1929), 180. Opinion as to the nature of James's achievement is divided. In his essay "A Majestic Literary Fossil," Mark Twain noted that the *Medicinal Dictionary* was published at the time of the rebellion of '45, and that "If it had been sent against the Pretender's troops there probably wouldn't have been a survivor." Hazen cites the opinion of Bartholomew Parr, recorded in his preface to the *London Medical Dictionary*. Parr notes James's extensive erudition, but appropriately faults him for diffuseness of language and a failure to provide pointed and decisive views, a necessity for a reference work which would be consulted in emergencies. See *Prefaces & Dedications*, p. 71.

53. See above, note 33.

the life of the Royal Society's historian-apologist, Bishop Sprat.

Many of the medical biographies are adapted or "translated" from intermediary sources. In the *Life of Boerhaave* Johnson used the work of Boerhaave's close friend, Albert Schultens, *Oratio academica in memoriam Hermanni Boerhaavii*; for "Morin," "Ruysch," and "Tournefort" he used Fontenelle's *Éloges*, and for "Aesculapius," "Archagathus," "Aretaeus," and "Asclepiades" he employed LeClerc's *Histoire de la Médecine*. Of course, the fact that he often depended on these sources for factual material in no way invalidates the resulting works as mines of information concerning his own attitudes. As has often been noted, when he appropriates material for his own use he at times "translates" (usually improving the style in the process), but he generally omits, expands, epitomizes, and, perhaps most important, editorializes.[54] His editorial comments and digressions often take the form of incisive and illuminating general statements; in most instances it can be shown that when Johnson draws upon one or more sources he truly makes the work in hand his own, and we can regard that work, with appropriate qualifications, as an original composition.

Johnson also wrote prefaces and dedications for scientific or quasi-scientific books. The Proposals and Dedication for the *Medicinal Dictionary* are his. In addition, he wrote the Preface to Dodsley's *Preceptor* (1748), a textbook and aid to self-education which contained sections on astronomy, geometry, and natural history;[55] the Dedication for George Adams' *Treatise on the Globes* (1766), and the Dedications for William Payne's *Introduction to Geometry* (1767) and *Elements of Trigonometry* (1772). His claims and appeals in these works, as would be expected, are sometimes inflated, but Boswell's statement that "in writing Dedications for others, he considered himself as by no means speaking his own sentiments" (*Life*, II, 2) is

54. On Johnson and "translation," see Abbott, "Dr. Johnson's Translations from the French"; Joel J. Gold, "Johnson's Translation of Lobo," *PMLA*, 80 (March 1965), 51–61; E. L. McAdam, Jr., "Johnson's Lives of Sarpi, Blake, and Drake," ibid., 58 (June 1943), 466–76.

55. See Roger P. McCutcheon, "Johnson's and Dodsley's *Preceptor*, 1748," *Tulane Studies in English*, 3 (1952), 125–32.

really not relevant in this case; the principles set down in these minor works square with statements made in other writings.

Interesting, often overlooked material is to be found in his reviews of scientific works in the *Literary Magazine* (for example, of Birch's history of the Royal Society, the Newton-Bentley correspondence, Home's experiments on bleaching, Stephen Hales's work on the distilling of sea water, ventilators in ships, and the curing of poor taste in milk, Lucas' essay on waters, and vol. XLIX, pt. 1 of the *Philosophical Transactions*) and in notices of scientific books in the *Gentleman's* such as Baker's "Employment for the microscope" or Feijóo's "Exposition of the uncertainties in the practice of physic." These pieces usually contain long excerpts from the works discussed and sometimes include only a paragraph or two of commentary, but their value could hardly be denied on the basis of brevity.

The abstractly definable "subject" of a Johnsonian work is often elusive. No *catalogue raisonné* of his "scientific writings" is really possible, for the majority of those works which chiefly treat science treat other things as well, and many works which are ostensibly non-scientific contain important comments on natural philosophy or the men who make science their occupation. Since "science's" ramifications are so wide and since Johnson draws on all of his experience whenever he writes, all of his works are potentially relevant. For Johnson—it has often been pointed out—human experience is of a piece. Scientific, philosophic, political, literary, historical, and religious principles are mutually illuminating. As he writes, the most ostensibly disparate themes may be interwoven. When he seeks to enforce a point he searches all of his reading and experience for examples. A discussion that could be limited to poetry soon shades into scientific matters; moral preoccupations can lead him into a treatment of history, politics, or philosophy. He is always flexibly comprehensive.

Within the works which are especially relevant to the present study, science can and often does form part of a larger discussion. In *Adventurer* 99, for example, he discusses "projectors," the planners of grandiose schemes which may end in success or ridicule. Men of science are among his examples, but in the course of the essay he also mentions Shakespeare's Coriolanus, Catiline, Xerxes, Alexander the Great, Peter the Great, Columbus, and Charles XII of Sweden. "Science" is submerged within greater issues. The process is often reversed. Just as scientific questions can find their place in larger discussions, Johnson can move from a treatment of science or scientists to other matters. In the *Life of Sydenham* he questions the "false report" that Sydenham plunged into the practice of medicine without a thorough study of his predecessors. He offers the following rationale for the perpetuation of this notion:

> But if it be, on the other part, remembered, how much this opinion favours the laziness of some, and the pride of others; how readily some men confide in natural sagacity; and how willingly most would spare themselves the labour of accurate reading and tedious inquiry; it will be easily discovered, how much the interest of multitudes was engaged in the production and continuance of this opinion, and how cheaply those, of whom it was known that they practised physick before they studied it, might satisfy themselves and others with the example of the illustrious Sydenham (1825 *Works*, VI, 409).

Here Johnson begins with a concrete occurrence and branches out through a discussion of human failings and human psychology to point out the common nature of the experience, and its relevance to any reader, even one who is only remotely interested in science, medicine, or Thomas Sydenham.

The pattern is evident in an essay such as *Idler* 24, a discussion of the important question raised by Descartes and Locke, "whether the soul always thinks?" Johnson examines the arguments which have been advanced and finds them wanting, but rather than leave the matter unresolved he offers "an argument, hitherto overlooked, which may perhaps, determine the controversy":

> If it be impossible to think without materials, there must necessarily be minds that do not always think; and whence shall we furnish materials for the meditation of the glutton between his meals, of the sportsman in a rainy month, of the annuitant between the days of quarterly payment, of the politician when the mails are detained by contrary winds.
> But how frequent soever may be the examples of existence with-

out thought, it is certainly a state not much to be desired. He that lives in torpid insensibility, wants nothing of a carcase but putrefaction. It is the part of every inhabitant of the earth to partake the pains and pleasures of his fellow beings; and, as in a road through a country desart and uniform, the traveller languishes for want of amusement; so the passage of life will be tedious and irksome to him who does not beguile it by diversified ideas (*Idler*, p. 77).

Johnson's argument does not constitute an "answer" so much as a raising of the point at issue to a higher plane. He writes as satiric moralist, assaulting the listless, the self-indulgent, the single-minded trifler. At base, the essay is an indictment of those who impoverish their lives and invite the representation of themselves as literary types. In other words, there are individuals whose range of interests is so narrow, whose sense of selflessness —both moral and intellectual—is so limited, that they can justly be sketched in simple, conventional outline. It is quite likely that Johnson is here reworking Locke and insisting "that it is at all times in a man's power to think or not to think," as Rodman D. Rhodes, in an important essay, suggests,[56] but Johnson's ultimate focus is articulated in the final sentence: the necessity of avoiding "torpid insensibility," the grinding tedium which besets the man who does not "partake the pains and pleasures of his fellow beings" through rigorous, diversified thinking.

Idler 24, it is true, deals with epistemology and not natural philosophy as such, but the two are, of course, closely related. The essay begins with a brief discussion of animals and birds and the question of whether or not their mental capabilities extend beyond what we would term instinct. The Idler's interest, however, is in men and so the discussion shifts. The tendency to begin with the natural world and suddenly turn to the moral is common; in other essays—as will be seen in Chapter IV—Johnson will consider the problem of diversifying one's ideas and suggest science or even scientific tinkering as a worthy pursuit.

The range of Johnson's interests and his ability to bring a variety of knowledge and experience to bear on a given subject are central aspects of his mind and art. The fact that he can support

56. "*Idler* no. 24 and Johnson's Epistemology," *Modern Philology*, 64 (Aug. 1966), 10–21.

a principle with wide and various examples or apply a single notion to very different areas of human life and thought are marks of his genius, but they also constitute an especially appropriate method of procedure for a man writing in popular forms. By and large the readers of Johnson's medical biographies, periodical essays, prefaces, dedications, and reviews would not desire specialized knowledge expressed in technical terms. By presenting his facts and commentary in terms of general human experience Johnson is able to popularize without resorting to the lesser man's practice of simply diluting complex material. As a result of his method Johnson's "writings on science" are eminently readable; with regard to subject or focus, however, they are often difficult to classify.

Professor Wimsatt has authoritatively demonstrated the omnipresent influence of scientific reading on Johnson's prose style: his fondness for both commonplace and recondite "philosophic" diction, the use of scientific similes, metaphors, and analogies for illustrative purposes, the "persistent process of metaphoric transfer from the realm of the philosophic to that of the psychological."[57] The reader who has never inspected Johnson's medical biographies, scattered scientific pieces, and pertinent essays could, with little difficulty, infer Johnson's continual interest in scientific matters through the sheer bulk of scientific references, allusions, and commonplaces in his work. He is so at ease with "philosophy" that he employs it as an entrée into other subjects; he relies on scientific principles or observations to facilitate the progress of his discussion, no matter what the subject of that discussion may be.[58] To be sure, passing references to science also enhanced his powers of conversation. When Boswell asserted that his father was content, and offered as evidence the fact that he was not restless, Johnson replied, "Sir, he is only locally at rest. A chymist is locally at rest; but his mind is hard at work" (*Life*, III, 241). And when Boswell claimed that many are happy living in the country, Johnson responded, "Sir, it is in the intellectual world as in the physical world; we are told by natural

57. *Philosophic Words*, p. 93.
58. See the beginnings of, e.g., *The False Alarm*, *Idler* 37, *Idler* 43.

philosophers that a body is at rest in the place that is fit for it; they who are content to live in the country, are *fit* for the country" (*Life*, IV, 338).

The perception of the use of science for the purpose of illustration, the weaving of "philosophy" into the texture of prose and verse, is central to an appreciation of the period's literature. Sprat had promised new material for literature; a receptive audience was not wanting. However, as Wordsworth indicated in a famous passage in the preface to the second edition of *Lyrical Ballads*, remote scientific discoveries can serve as proper material for the literary artist only when "these things shall be familiar to us, and the relations under which they are contemplated by the followers of these . . . sciences shall be manifestly and palpably material to us as enjoying and suffering beings."[59] Johnson accomplishes this integration. His predecessors and contemporaries very often do not, and the presence of scientific rhetoric in the works of minor versifiers often strikes a modern reader as awkward and unintentionally ludicrous. In this connection, claims that the scientific revolutions materially improved the level of work of second-line poetic talents, that "philosophic" material enabled poets to rise above the conventional and represent the sublime, should be viewed with a healthy portion of skeptical caution. By the same token, Johnson's scientific rhetoric deserves even more attention than it has already been accorded. Close comparison of Johnson's practices—as outlined by Professor Wimsatt, but with considerably more stress placed on imagery[60] —with those of other prose writers would, I think, enhance the stature of his accomplishment because of the completeness of the integration of scientific material with all of his experience and the attending facility and clarity of its communication.

59. *The Poetical Works of Wordsworth*, ed. Thomas Hutchinson, rev. Ernest de Selincourt (London, 1951), p. 738.

60. See Donald J. Greene, " 'Pictures to the Mind': Johnson and Imagery," *Johnson, Boswell and their Circle: Essays Presented to Lawrence Fitzroy Powell in Honour of his Eighty-Fourth Birthday* (Oxford, 1965), p. 158, n. 2.

CHAPTER III

The Baconian Legacy

Little energy need be expended in demonstrating the fact that Newton was not only the model scientist for the eighteenth century but the model thinker as well. Wordsworth's lines in Book III of *The Prelude* ("the statue . . . /Of Newton with his prism and silent face,/The marble index of a mind for ever/ Voyaging through strange seas of Thought, alone" ll. 60–63) are perhaps the most famous poetic treatment of Newton, but the reverent awe of Halley's ode, published with the *Principia*, is more characteristic of the age's response:

> Talia monstrantem mecum celebrate Camaenis,
> Vos qui coelesti gaudetis nectare vesci,
> NEWTONUM clausi reserantem scrinia Veri,
> NEWTONUM Musis charum, cui pectore puro
> Phoebus adest, totoque incessit Numine mentem:
> Nec fas est propius Mortali attingere Divos.[1]

1. O ye who take joy in taking heaven's nectar, celebrate with me in song the one who shows such things: Newton who unlocks the chest of imprisoned truth: Newton, to the muses dear. Phoebus is in his pure breast;

59

The success of Newton did not, however, obscure the ideological importance of Bacon. Newton had come to fulfill, not to replace. As Horace Walpole stated, Bacon was "THE PROPHET OF ARTS, which Newton was sent afterwards to reveal."[2] The biblical terms are commonplace; like Cowley, Thomson compared Bacon to Moses:

> The great Deliverer he! who from the Gloom
> Of cloyster'd Monks, and Jargon-teaching Schools,
> Led forth the true Philosophy, there long
> Held in the magic Chain of Words and Forms,
> And Definitions void: he led Her forth,
> Daughter of HEAVEN! that, slow-ascending still,
> Investigating sure the Chain of Things,
> With radiant Finger points to HEAVEN again.
>
> (*Summer*, 1528–35)

Goldsmith, learning heavily on the *Encyclopédie*, noted that Bacon "first discovered the general principles, which were to serve as guides in the study of Nature. He first proposed the usefulness of experiments alone, and hinted at several, which others afterwards made with success." "Newton," Goldsmith wrote, "by uniting experiments with mathematical calculations, discovered new laws of Nature, in a manner at once precise, profound, and amazing." The sum total of the contributions of Newton's predecessors, Goldsmith argued, "does not amount to the tenth part of the discoveries of the *English* philosopher only."[3]

The articles of Bacon's faith are revolutionary but few in number. In a sense the strength of his ideological influence turns on the fact that his program is so clearly outlined. Since it enjoys the beauty of simplicity it can be developed or extended without being supplanted; it can pass into legend without being superseded.

---

he has entered his mind with total divinity. Nearer the gods no mortal may approach.

2. *A Catalogue of the Royal and Noble Authors of England, with Lists of their Works*, new ed. (London, 1796), p. 128.

3. "Introduction" to *A Survey of Experimental Philosophy*, in *Collected Works of Oliver Goldsmith*, ed. Arthur Friedman, v (Oxford, 1966), 344, 346.

Thus, Newton can reach a level of achievement beyond Bacon's scientific capabilities, but remain within the broad outline which constitutes the Baconian tradition. The chief difference, for example, between the method which Newton practiced and the method which Bacon advocated concerns the premium which Newton and his mentor Barrow placed on mathematics. (Bacon's relative ignorance of mathematics is not so surprising as Whig history would suggest. Pepys, for example, did not know simple arithmetic and was unable to multiply or divide. He did not begin to study mathematics until 1662. John Wallis, perhaps the most distinguished English mathematician before Newton, had not even heard of arithmetic until, at about the age of fifteen, his brother lent him the textbook he was studying.)[4] Newton's method, however, represented development rather than departure. What he and Barrow did was to combine Baconian experimentation with a mathematical and quantitative technique which enabled them to interpret Bacon in a new, more sophisticated fashion. The resulting synthesis combined the rigor of mathematics with the certainty of experiment.[5] Unfortunately, it brought criticism before adulation. In setting forth the quantitative laws of motion and their consequences in the *Principia*, Newton was offering an abstract, mathematical description in place of the pictorial depictions favored by the mechanical philosophers. Hence the initial response on the continent was that Newton's work was a dazzling display of mathematics, but not physics at all. Eventually such writers as d'Alembert, Voltaire, and Condillac would be by and large agreed that physical science at its best can only be a mathematical description of the laws of nature, but favorable response was not immediately forthcoming.[6]

This distinction between the Baconian and Newtonian meth-

---

4. Marjorie Nicolson, *Pepys' "Diary" and the New Science* (Charlottesville, 1965), pp. 8–9.

5. Robert Hugh Kargon, *Atomism in England From Hariot to Newton* (Oxford, 1966), pp. 118–19, 121.

6. See Henry Guerlac, "Where the Statue Stood: Divergent Loyalties to Newton in the Eighteenth Century," in *Aspects of the Eighteenth Century*, ed. Earl R. Wasserman (Baltimore, 1965), pp. 317–34.

odologies concerning the importance of mathematics is extremely important, but Bacon's program for the advance of natural
philosophy includes far more than the method alone. In enlarging upon Bacon, Newton did not jettison the inductive method,
in Bacon's view the *sine qua non*—the crucial factor differentiating the new philosophy from the old—and did not significantly
disturb such ancillary Baconian principles as the attack on en-
slavement to "authority," the importance of utilitarian applications of pure science, or the necessity of deeds as opposed to
words. In Johnson's century, experimental, as opposed to mechanical, philosophy was continually praised; Bacon is its progenitor, Newton its major practitioner. As will be seen, Johnson
is in essential agreement with the outline of Bacon's program,
and, ideologically, English science is Baconian science.

Early in the *Great Instauration*, Bacon indicates that the
"greatest change" from previous scientific practice which he is
introducing "is in the form itself of induction and the judgment
made thereby." But, he hastens to add, the induction "which
proceeds by simple enumeration, is a puerile thing; concludes at
hazard; is always liable to be upset by a contradictory instance;
takes into account only what is known and ordinary; and leads
to no result."[7] As Joseph A. Mazzeo points out, in an excellent
summary discussion of Bacon's program and influence,[8] Bacon's
ideas were often distorted: mindless experimentation, brute empiricism was carried on under the banner of Baconian science.
But to Bacon the "empiric" is like the ant, amassing mountains of
material which will go unaltered. The scholastic rationalist is the
spider spinning webs out of his own entrails, while the model is
the bee, combining empiricism and rationalism, gathering substance from various sources, adding his own material to it, and,
finally, producing honey. Thus, when Johnson defines "empiricism" in the *Dictionary* as "Dependence on experience without

7. *The Works of Francis Bacon*, ed. J. Spedding, R. L. Ellis, and D. D.
Heath, IV (London, 1857–74), 25. All subsequent references are to this
edition.
8. *Renaissance and Revolution: Backgrounds to Seventeenth-Century
English Literature* (New York, 1967), pp. 204–5.

knowledge or art; quackery," his position is that of Bacon. Terminology is important; Johnson approves of and encourages induction and the experimental philosophy, but the ways of the "empiric" are criticized in his works, just as they are in the *Advancement of Learning* and *Novum Organum*.

An appropriate remark in Johnson's review of Dr. Charles Lucas' *Essay on Waters* will illustrate the point. Lucas has taken two glass vials, one containing a drop of water, the other filled with gunpowder. Lucas stops them, heats them, and notices that the vial containing the water bursts with greater noise and force. He concludes that "*the power of rarified water is greater than that of inflamed gunpowder.*" Johnson says he will not deny the "position infered" but he does not think that it follows from the experiment. He provides his reason and then comments: "So useless are these trials which an elegant writer has lately degraded to their proper rank by the name of *bruta experientia*, unless theory brings her light to direct their application."[9]

The blind experimentation of the empiric is likely to lead to fallacious conclusions, since it is not guided by principle or working hypothesis. One cannot merely begin experimenting and hope to arrive at valid conclusions. If the context of one's experiments is not firmly established, a lapse into absurdity is likely. Experimentation without learning is as useless as knowledge gained from books which is never put to the empirical test. Study and experiment interact, and experiment is both guided by and rises to principle. Johnson is by no means suggesting that experimentation should be shackled by "hypotheses," a term whose importance—because of Newton's "hypotheses non fingo" —is equalled only by its ambiguity. If, as Henry Guerlac suggests, a hypothesis for Newton is an a priori assumption not scrupulously inferred from experience,[10] then Johnson wants no part of it. The "directing theory" for Johnson must be based on

9. *Literary Magazine,* 1, iv, 168.
10. "Where the Statue Stood," p. 327. For an excellent discussion of the possible meanings intended by Newton in his famous dictum, see Maurice Mandelbaum, *Philosophy, Science, and Sense Perception: Historical and Critical Studies* (Baltimore, 1966), pp. 71–77.

or deduced from observed phenomena. As has often been noted, Johnson is wary of systems and systematizers; the systematizer's stock in trade is, in Newton's sense of the word, the hypothesis. Elsewhere, Johnson will dissuade the would-be philosopher from following the high Priori Road; here he is making the thoroughly Baconian point that undisciplined empiricism—which Bacon associated with the alchemists—infects the imagination and brings no useful or credible result. Because of human fallibility, Bacon advises the close control of experimentation; reason must aid the all too easily deceived human senses. In a lighter vein, Johnson comments on Lucas' assertion that an old man placed in a certain cold bath will enjoy "an uninterrupted state of health": "This instance does not prove that the cold bath produces health, but only, that it will not always destroy it. He is well with the bath, he would have been well without it."[11] The flavor of the remark is delightful, but the quite serious principle is being reapplied.

Sprat's discussion of the Royal Society's skeptical cast of mind has already been mentioned. Sprat sees the circumspect, tentative approach as a middle road between dogmatism and crippling skepticism. This is, of course, Baconian. In the preface to the *Novum Organum* Bacon distinguishes between those who believe they are in firm possession of the laws of nature and those who feel that absolutely nothing can be known concerning those laws. Both extremes, presumption on the one hand and despair on the other, stifle inquiry, so that Bacon offers an alternative approach involving progressive stages of certainty. For Johnson also, "skepticism" is an extreme, an intellectual stance which automatically precludes investigation. He defines it in the *Dictionary* as "Universal doubt; [the] pretence or profession of universal doubt." In a book notice for the *Gentleman's Magazine* of Benito Geronimo Feijóo's "An exposition of the uncertainties in the practice of physic," he writes:

11. *Literary Magazine*, 1, v, 229. One cannot resist quoting a further remark (p. 229): "The rules about the posture to be used in the bath, and the directions to forbear to speak during the action of the water, are refinements too minute to deserve attention, he is past much hope from baths to whom speech or silence can make any difference."

This treatise is intended to shew the total inefficacy of physick for the restoration of health. . . . By this work he [Feijóo] appears to have great abilities, yet he seems to have display'd them rather in favour of scepticism than truth. The effect of medicines with respect to the cure of particular diseases, is indeed in a great degree uncertain, and they are frequently apply'd without success, because the disease is not sufficiently known, and the circumstances of the patient with respect to situation, habit, manner of life, and constitution are not regarded with sufficient attention. But tho' medicines are sometimes applied without success, the effects of many are known and certain . . . .[12]

To quote from Johnson's definition of "incredulous," he himself is "hard of belief." His perception of human frailty, of man's penchant for error and half-truth, informs all of his thought, but because of his awareness of human failings he is extremely sympathetic to all of man's successes. Academic and Pyrrhonian skepticism represent, to Johnson, capitulation.[13]

Intellectual triumphs are frequently the result of the rigorous

12. *Gentleman's Magazine*, 21 (Feb. 1751), 95. Attributed to Johnson by Arthur Sherbo, "Samuel Johnson and the *Gentleman's Magazine*, 1750–1755," in *Johnsonian Studies*, ed. Magdi Wahba (Cairo, 1962), p. 139. It should be noted that whenever I cite works which have been attributed since the Courtney-Smith bibliography, I am recording concurrence with the attribution.

13. Academic skepticism, which dominated the philosophy of the Platonic Academy until the first century before Christ, held that no knowledge is possible. Pyrrhonian skepticism, connected primarily with the writings of Sextus Empiricus (ca. A.D. 200), held that there is insufficient and inadequate evidence for determining if any knowledge is possible. Hence, one ought to suspend judgment on all questions concerning knowledge. See Richard H. Popkin, *The History of Scepticism from Erasmus to Descartes*, rev. ed. (New York, 1964), pp. ix–xi. Johnson eschews the title of skeptic, and is not, by any stretch of the imagination, a Pyrrhonist. His "skepticism," or rather "incredulity," should be associated with the practices and pronouncements of the Royal Society. Phillip Harth has fully outlined this scientific circumspection and diffidence. See his *Contexts of Dryden's Thought* (Chicago, 1968), pp. 1–31. The necessity of following Harth's practice and carefully discriminating the varieties of seventeenth- and eighteenth-century skepticism cannot be overstressed. The question of Johnson's attitude toward Pyrrhonism recently arose in the Greene-Stanlis interchange concerning Johnson and

application of the method of experimental philosophy. The most cursory glance at Johnson's work reveals his preference for direct observation and experience over theories and speculative systems. "Every day," he writes in *The False Alarm*, we see "the towering head of speculation bow down unwillingly to groveling experience."[14] "Human experience, which is constantly contradicting theory, is the great test of truth"(*Life*, I, 454). "The foundation [of knowledge] ... must be laid by reading. General principles must be had from books, which, however, must be brought to the test of real life"(*Life*, II, 361). The astronomer in *Rasselas* considers his control of the weather and seasons incredible for the very important reason that he "cannot prove it by any external evidence ...."[15]

The empirical test extends to all areas of human experience. It is central to scientific activity, but reliance on one's own perception and the accuracy which results from first-hand observation are guiding principles for all who would instruct or influence others. Johnson reserves special praise for those who see for themselves and convey their experience and observations accurately. Father Lobo "appears, by his modest and unaffected narration, to have described things as he saw them, to have copied nature from the life, and to have consulted his senses, not his imagination."[16] Shakespeare, "whether life or nature be his subject, shews plainly, that he has seen with his own eyes ... the ignorant feel his representations to be just, and the learned see that they are compleat."[17] Johnson prefers Boswell's Journal to his History in the *Account of Corsica*: "Your history was copied from books; your journal rose out of your own experience and

---

"Natural Law" in the *Journal of British Studies*, 2–3 (1963), 86, 162, 166–67. Greene raises the matter again in his *Samuel Johnson: A Collection of Critical Essays* (Englewood Cliffs, 1965), p. 7.

14. 1825 *Works*, VI, 164.

15. *Johnson: The History of Rasselas, Prince of Abissinia*, ed. J. P. Hardy (Oxford, 1968), p. 102. [Cited hereafter in text and notes as *Rasselas*.]

16. 1825 *Works*, V (*Preface to the Translation of Father Lobo's Voyage to Abyssinia*), 255.

17. *Johnson on Shakespeare*, ed. Arthur Sherbo, 2 vols. (New Haven, 1968), VII, 89–90.

observation."[18] During his last illness he praised Gillespie's medical advice as "solid practical experimental knowledge."[19]

Others are advised to trust to the knowledge attained by the senses. In 1762 Johnson writes to Dr. George Staunton:

In America there is little to be observed except natural curiosities. The new world must have many vegetables and animals with which philosophers are but little acquainted. I hope you will furnish yourself with some books of natural history, and some glasses and other instruments of observation. Trust as little as you can to report; examine all you can by your own senses. I do not doubt but you will be able to add much to knowledge, and, perhaps, to medicine.[20]

He gives similar advice to Susannah Thrale:

I am glad that you and your sisters have been at Portland. You now can tell what is a quarry, and what is a cliff. Take all opportunities of filling your mind with genuine scenes of nature. Description is always fallacious, at least till you have seen realities, you can not know it to be true.[21]

The scientific methodology which Johnson recommends is, of course, based on experiment and observation, and consists of three elements. After a study of the works of his predecessors and contemporaries, the scientist examines nature for himself, decides whether the voice of authority should be heeded or discarded, and finally moves from his own observations to principles. Study provides the philosopher a context within which to work; his own empirical activities enable him to judge matters for himself and move to new discoveries, while his "reason" leads him to synthesize his observations into axioms and principles. All three elements are necessary. The study of books may prove inhibiting but it is nevertheless indispensable. For science to progress, the empirical analysis which has been guided—to a greater or lesser extent—by knowledge gained from books requires synthesis. Empiricism is not an end in itself: "reason," "good sense," or "judgment" (Johnson uses all three terms) must enable the

18. *Letters*, I, no. 222 (9 Sept. 1769 to Boswell), 230.
19. *Letters*, III, no. 937 (2 March 1784 to Boswell), 137.
20. *Letters*, I, no. 140 (1 June 1762), 136–37.
21. *Letters*, III, no. 880 (9 Sept. 1783), 69.

scientist to generalize after his observations are completed. Henry Baker's "Employment for the microscope" is praised for facilitating this procedure: "By this book a new field is opened to curiosity, and a great variety of facts are established, as data for new operations of judgment."[22]

Like Bacon and his seventeenth-century followers, Johnson tends to disparage scientific studies conducted in libraries rather than laboratories. He says of the Greek physician Alexander: "He appears through his whole Works to have attended diligently not only the Instructions of his Predecessors, but to the Precepts of far greater Certainty, the Dictates of Reason, and the Evidence of Experience."[23] Bacon realizes that an investigator cannot begin any study without a notion of the framework in which he is to operate, but he has little patience with "authors in sciences" who have been revered as dictators rather than advisors and thus curtailed inquiry,[24] for "disciples do owe unto masters only a temporary belief and a suspension of their own judgment until they be fully instructed, and not an absolute resignation or perpetual captivity . . . ."[25] Johnson would also counsel philosophers to place greater faith in their own abilities and distrust the accounts of others. As Nekayah says in *Rasselas*, "What reason cannot collect . . . and what experiment has not yet taught, can be known only from the report of others" (*Rasselas*, p. 70). "Report" is the recourse of the novice.

Johnson's fullest exposition of the scientist's methodology as he conceives it occurs in the *Life of Boerhaave*. Boerhaave's empiricism receives extensive description: we see him inspecting the bodies of different animals, examining the plants in the garden of the university, and making excursions into the woods and fields. Johnson approves of Boerhaave's practices:

22. Book notice, *Gentleman's Magazine*, 23 (May 1753), 251. Attributed by Sherbo, "Samuel Johnson and the *Gentleman's Magazine*, 1750–1755," p. 145.
23. "Alexander," *Med. Dict.*, ı, sig. Sssl^v.
24. Bacon, *Works*, ııı (*The Proficience and Advancement of Learning*), 289.
25. Ibid., p. 290.

When he laid down his Office of Governor of the University in 1715. he made an Oration upon the Subject of *attaining to Certainty in Natural Philosophy*; in which he declares himself, in the strongest Terms, a Favourer of experimental Knowledge, and reflects with just Severity upon those arrogant Philosophers, who are too easily disgusted with the slow Methods of obtaining true Notions by frequent Experiments, and who, possess'd with too high an Opinion of their own Abilities, rather chuse to consult their own Imaginations, than inquire into Nature; and are better pleased with the delightful Amusement of forming Hypotheses, than the toilsome Drudgery of amassing Observations.

The Emptiness and Uncertainty of all those Systems, whether venerable for their Antiquity, or agreeable for their Novelty, he has evidently shewn; and not only declar'd, but prov'd, that we are entirely ignorant of the Principles of Things, and that all the Knowledge we have is of such Qualities alone as are discoverable by Experience, or such as may be deduced from them by Mathematical Demonstration.[26]

Boerhaave's empiricism is not an unsystematic, blind collection of data. He has studied the Ancients as well as the Moderns. His method is ordered, and his acquaintance with the works of his predecessors and contemporaries adds stability and solidity to the observational aspects of his investigations:

He examined Systems by Experiments, and formed Experiments into Systems. He neither neglected the Observations of others, nor blindly submitted to celebrated Names. He neither thought so highly of himself, as to imagine he could receive no Light from Books, nor so meanly, as to believe he could discover nothing but what was to be learned from them. He examined the Observations of other Men, but trusted only to his own.[27]

The empiricism is central, study of others necessary, and the ascent to principles the end result:

This [Boerhaave's public lectures on chemistry] he undertook, not only to the great Advantage of his Pupils, but to the great Improvement of the Art itself, which had been hitherto treated only in a

26. "Boerhaave," *Med. Dict.*, I, sig. 9U2r.
27. Ibid., sig. 9X1r.

confus'd and irregular manner, and was little more than a History of particular Experiments, not reduced to certain Principles, nor connected one with another: This vast Chaos he reduced to Order, and made that clear and easy, which was before to the last degree perplex'd and obscure.[28]

All of the elements in Johnson's methodology are here in the works and principles of Boerhaave: the ordered study, diligent observation and experimentation, and the application of reason to data which results in synthesis. Two balances are carefully maintained, that between study and personal observation, and that between analysis (the examination of data and previous findings) and synthesis. The method should be kept in mind. Because Johnson tends to distrust the studies of others and places little faith in systems and hypotheses—at worst because they are self-spun philosophical "romances," at best because they are premature—he is inclined to treat the experimental element as if it is all important, and not simply the foundation of the method. His assaults, for example, on "tradition" and "authority" can lead to the mistaken notion that he is encouraging scientific self-sufficiency.

In *Rambler* 154 Johnson claims that "the mental disease of the present generation, is impatience of study, contempt of the great masters of ancient wisdom, and a disposition to rely wholly upon unassisted genius and natural sagacity."[29] Hurried pride is folly. One can "discover" something that has long been known, as Barretier did in his studies of the longitude,[30] and "waste, in attempts which have already succeeded or miscarried, that time which might have been spent with usefulness and honour upon new undertakings" (*Rambler*, v, no. 154, 58). Every student, scientist, or poet must study his predecessors, but Johnson warns against excessive regard for authority and is particularly pleased when the student of science boldly flouts tradition and strikes out on his own. In discussing Boerhaave's *Indexes* he writes:

28. Ibid., sig. 9U2ʳ.
29. *Samuel Johnson: The Rambler*, ed. W. J. Bate and Albrecht B. Strauss, 3 vols. (New Haven, 1969), v, 55. [Cited hereafter in text and notes as *Rambler*.]
30. 1825 *Works*, vi (*Life of Barretier*), 389.

In the Conduct of his Botanical Works, he has discover'd a Mind open to Truth, and entirely free from that base and servile Attachment to *Names* and *Authority*, which has, in all Ages, prov'd the Bane of Learning and good Sense. He had a Judgment of his own, and bravely dar'd to use it.[31]

In the *Life of Sydenham* he admits he has no information concerning Sydenham's childhood, but he cannot resist a general comment:

We must, therefore, repress that curiosity, which would naturally incline us to watch the first attempts of so vigorous a mind, to pursue it in its childish inquiries, and see it struggling with rustick prejudices, breaking, on trifling occasions, the shackles of credulity, and giving proofs, in its casual excursions, that it was formed to shake off the yoke of prescription, and dispel the phantoms of hypothesis (1825 *Works*, vi, 406).

Later in the biography, Johnson mentions Sydenham's "contempt of pernicious methods, supported only by authority, in opposition to sound reason and indubitable experience."[32] His approval of scientific debunking is evident in his praise of Sir Thomas Browne's *Pseudodoxia Epidemica*, for which he wishes a new supplement: "It might now be proper . . . to reprint it with notes, partly supplemental, and partly emendatory, to subjoin those discoveries which the industry of the last age has made, and correct those mistakes which the author has committed, not by idleness or negligence, but for want of Boyle's and Newton's philosophy."[33]

If Johnson tends to minimize the importance of tradition and authority, the systematizer receives even less mercy. His remarks on Actuarius in the *Medicinal Dictionary* constitute a kind of thesis statement and leave little doubt concerning his position:

He had a great Propension to Theory and Ratiocination, but was not contented to form Systems in his Closet, but extended his Speculations to Distempers and Symptoms with which he was only ac-

31. "Botany," *Med. Dict.*, I, sig. 10E1r.
32. 1825 *Works*, vi, 411.
33. Ibid. (*Life of Browne*), p. 483.

quainted by the Means of Books, which have always been found fallacious and uncertain Guides . . . . It was my Opinion, says he, that Methods of Cure, not founded upon Reasoning, never could be relied on; and that a just Theory would make Physic not only a more easy Study, but a more successful Profession.

As the Authority of *Actuarius* is not sufficiently established, to mislead any of our Readers, it is not necessary to separate with great Accuracy his Errors from his just Notions. I shall only observe, that Theory may make Physic easy, but its Success must arise from Experience.[34]

Far more sympathy is expressed for the practices of a man like Tournefort, the French botanist and physician, who, at the beginning of his studies, "discover'd no great Relish for what was taught him" and found, in philosophical disquisitions, "only vague and abstracted Ideas, which decoy and amuse the Mind, without enriching it with any thing that is solid and satisfactory."[35] Johnson enjoys writing for victory of course, and since such issues as the assault on system and authority are ideological commonplaces, we must be careful to temper such comments as have just been quoted with more balanced pronouncements. Nevertheless, the Johnson who comments on natural philosophy is the same Johnson who feels "that all rebellion [is] natural to man" (*Tour*, p. 385)—as far removed as possible from the mythological construct who vigorously defends the *status quo*, and reveres tradition to the extent that he bypasses the truth in his haste to oppose change and progressive development.

In the prefatory paragraph to the *Gentleman's Magazine* version of the *Life of Boerhaave*, Johnson indicates that the careful procedure of the scientist may also be employed by the scientist's biographer: "We could have made [the biography] much larger, by adopting flying Reports, and inserting unattested Facts; a close Adherence to Certainty has contracted our Narrative, and hindred it from swelling to that Bulk, at which modern Histories generally arrive."[36] Similar incredulity is exhibited in the *Life of*

34. "Actuarius," *Med. Dict.*, I, sig. Ii2^r.
35. "Botany," ibid., sig. 10C1^v.
36. *Gentleman's Magazine*, 9 (Jan. 1739), 37.

Sydenham. Numerous "false reports" had arisen and been maintained concerning Sydenham's career. They were either patently false, exaggerated, or misinterpreted, and Johnson announces that he will separate truth from falsehood. He enters into the spirit of the methodology which he approves and recommends, and, as biographer, assumes the role of the debunking skeptic, allowing nothing to pass that has not been established as fact and permitting no hypothetical reconstructions. Nearly 40 percent of what is in itself a very brief biography is devoted to these "reports."

However, the most interesting example of Johnson's application of the methods of the new philosophy appears in the *Journey to the Western Islands of Scotland.* Here he is working in an extremely practical context, testing accounts and relating the evidence of the senses. The processes depicted in the *Journey* constitute the best corrective to the one theme in Johnson studies which has most consistently impeded the understanding of Johnson's mind and art: the assumption that his intellectual life is uniformly stagnant. The popular images of Johnson, Macaulay's caricature and Boswell's sentimental portrait, are equally static, and Johnson's very rhetorical facility tends to mask the fact that his final pronouncements are the result of rigorous mental struggle. Moreover, he has not, as I noted earlier, been accorded the attention due a major figure[37] and many readers (including some commentators) are familiar only with the anthology pieces—the periodical essays, *Rasselas*, the major poems and chief *Lives*—where we see the distillation of experience, the products of a life of observation and careful discrimination. The developing nature of Johnson's thought, a focus on process rather than prod-

37. Donald Greene has pointed out the manner in which modern commentators often attempt summary pronouncements with little apparent knowledge of the body of Johnson's work. To generalize with only a handful of writings as basis reveals the flimsiest of methodologies. No self-respecting Miltonist, for example, would ever attempt such a procedure. See Greene, "The Development of the Johnson Canon," in *Restoration and Eighteenth-Century Literature: Essays in Honor of Alan Dugald McKillop*, ed. Carroll Camden (Chicago, 1963), pp. 407–8.

uct, is lost. Johnson's "reflections" are an important part of the *Journey*, as Arthur Sherbo has pointed out,[38] but the transition from empirical recording to discursive reflection is there traced in unique detail, and is, in a sense, one of the main subjects of the book. I am not, of course, suggesting that the *Journey* is the only work in which Johnson makes clear the facts or events which lead to general comment. In his biographies he often draws general conclusions after considering specific traits of character or actions in particular situations, but the process is especially clear in the *Journey* because of the fact that it resembles, in so many passages, a finished and refined diary. The immediacy of the experience is always evident. The Johnson of the *Journey* differs essentially from the counterfeit images of popular tradition. This work illustrates Johnson's intellectual temper so clearly that I single it out at this point for particular discussion.

Traveling between Anoch and Auchnasheal, Johnson and Boswell dismount in a narrow valley, the place, Johnson writes, where the initial conception of the *Journey* was formed:

> I sat down on a bank, such as a writer of Romance might have delighted to feign. I had indeed no trees to whisper over my head, but a clear rivulet streamed at my feet. The day was calm, the air soft, and all was rudeness, silence, and solitude. Before me, and on either side, were high hills, which by hindering the eye from ranging, forced the mind to find entertainment for itself. Whether I spent the hour well I know not; for here I first conceived the thought of this narration (*Journey*, p. 30).

In its subtle ways this important passage establishes the tone and spirit of the *Journey*. The traveler's eye is limited by the High-

38. "Johnson's Intent in the *Journey to the Western Islands of Scotland*," *Essays in Criticism*, 16 (Oct. 1966), 385–86. Sherbo's article represents one form of culmination to the controversy initiated by Jeffrey Hart. See Hart, "Johnson's *A Journey to the Western Islands*: History as Art," ibid., 10 (Jan. 1960), 44–59; Donald J. Greene, "Johnsonian Critics," ibid. (Oct. 1960), pp. 476–80; R. K. Kaul, "*A Journey to the Western Isles* Reconsidered," ibid., 13 (Oct. 1963), 341–50. See also Francis R. Hart, "Johnson as Philosophic Traveler: The Perfecting of an Idea," *ELH*, 36 (Dec. 1969), 679–95. There are several important points of contact between the last article and my "Johnson's *Journey*," *Journal of English and Germanic Philology*, 69 (April 1970), 292–303.

land topography and the result is a turning within, a recourse to the mind when the sight is blocked, the wide prospect and immediate view impossible. Johnson continually confronted "high hills," and though they took the form of superstition, the fantastic accounts of earlier historians coupled with a cavalier attitude toward detail, a national pride, or the pressure of common, untested opinion, the result was the same. The view is hampered; the observer is forced upon his own devices. Johnson must speculate, test, debunk, and imaginatively reconstruct. He takes pains to point out the manner in which his views are formed: the difficulties encountered, the sifting of evidence and clarification of experience, the concrete examples upon which his generalized reflections are based. To an extent, the book represents a mental diary, a history of the complexities of the learning process.

The observation of particulars, the necessary step which must precede and form the basis for reflections, is a difficult process and Johnson does not hesitate to portray the problems which beset his endeavor. Struggle is mirrored in style. A smooth and ordered presentation is sacrificed in the interest of factual completeness. Admissions of personal limitations, which could easily have been glossed over, appear frequently: "Men skilled in architecture might do what we did not attempt . . ." (p. 9). "To collect sufficient testimonies for the satisfaction of the publick, or of ourselves, would have required more time than we could bestow" (p. 82). "Mr. *Pennant*'s delineations, which are doubtless exact, have made my unskillful description less necessary" (p. 111). "Of Fort George I shall not attempt to give any account. I cannot delineate it scientifically . . ." (p. 19).

The style is fragmented. Paragraphing is often erratic. At many points the sentences and paragraphs end inconclusively. Johnson admits omissions of details or explanations in his account, despite the resulting choppiness:

We stopped a while at Dundee, *where I remember nothing remarkable*, and mounting our chaise again, came about the close of the day to Aberbrothick (p. 8).

At night we came to Bamff, *where I remember nothing that particularly claimed my attention* (p. 16).

How beasts of prey came into any islands is not easy to guess. In cold countries they take advantage of hard winters, and travel over the ice: *but this is a very scanty solution; for they are found where they have no discoverable means of coming* (p. 46).

Having wandered over those extensive plains, we committed ourselves again to the winds and waters; and after a voyage of about ten minutes, *in which we met with nothing very observable*, were again safe upon dry ground (p. 108).[39]

Transitions between paragraphs are often rough and Johnson's periods are frequently interrupted by curt interjections and weakened, ostensibly, by the tacking on of apparently irrelevant details. But rather than presenting a hastily written work, I believe that Johnson is suggesting, by means of stylistic devices and rhetorical ploys, the very process of moving from perception to reflection. The initial steps in the procedure are contrasted with the result and the grand generalizations are thus strengthened. At the close of the book Johnson comments, "Such are the things which this journey has given me an opportunity of seeing, and such are the reflections which that sight has raised." But in the course of his work he has indicated the fact that certain things were not seen, that some which were seen have been forgotten, that some which were seen were not justly described or evaluated. The simple balance between "things" and reflections suggested in his closing paragraph is really not simple at all and one of the chief themes of the *Journey* is the portrayal of the complexity of that balance.

One of the prime obstacles confronting Johnson in his journey is the uncertainty of common opinion. His faith in common assent is great. He is normally willing to accept the existence of a phenomenon if the body of common testimony is extensive; in his criticism, universal acceptance of a literary work is a reasonably reliable index to its quality. The systematizing philosopher, the literary critic with an exaggerated faith in his personal tenets, and the coterie, are all, at least initially, distrusted. Johnson's lot is cast with the general as opposed to the individual

39. Italics added throughout.

or esoteric. He is sympathetic, in other words, to the eighteenth-century norm which Lovejoy termed "uniformitarianism." Certain qualifications, however, must be made. One can either trust or distrust "common opinion," for it takes two different forms. The voice of "mankind" is more reliable than the special pleading of a systematizer or coterie. On the other hand, "mankind," as Bacon was quick to point out, is notoriously fallible and subject to particular types of deception. Johnson may hesitate to deny a commonly accepted belief or an experience known by all men, but he is well aware that "tradition," routine, and superstition can perpetuate falsehood with considerable ease. He is temperamentally inclined to take the side of "mankind" against the lone individual but whenever possible subjects common opinion to the empirical test. He would, needless to say, abhor the association of his initial trust in common testimony with credulity.

In the process of his travels Johnson is frustrated by the fact that the Scots assert boldly but differ widely in their accounts of the same event or practice. The stern statement leads to confusion and contradiction upon further questioning:

> He that travels in the Highlands may easily saturate his soul with intelligence, if he will acquiesce in the first account. The Highlander gives to every question an answer so prompt and peremptory, that skepticism itself is dared into silence, and the mind sinks before the bold reporter in unresisting credulity; but, if a second question be ventured, it breaks the enchantment; for it is immediately discovered, that what was told so confidently was told at hazard, and that such fearlessness of assertion was either the sport of negligence, or the refuge of ignorance.[40]

Johnson not only records his perturbation but throughout provides reflections based upon concrete incidents. In the *Rambler* he might have written that, "Accuracy of narration is not very common, and there are few so rigidly philosophical, as not to represent as perpetual, what is only frequent, or as constant, what is really casual." But in the *Journey* that comment

40. *Journey*, p. 38. Cf. pp. 83, 87–88.

is preceded by a discussion of the fact that "*Lough Ness* is open in the hardest winters, though a lake not far from it is covered with ice" (p. 22). What the Scots have accepted as delightfully strange and inexplicable, Johnson attempts to explain, and includes, in the process, a reflection based upon the particular case in point.

The Scots are not only inclined to negligence and the victims of ignorance, but also particularly protective concerning tradition and the national character. Their "patriotick vanity," as Johnson terms it, forces even a learned man like the Reverend Donald Macqueen of Skye to argue for the authenticity of the Ossian poems when he, in all probability, does not himself believe. Moreover, the accounts of previous historians are often unreliable. Martin's *Description of the Western Isles of Scotland* which accompanied Johnson and Boswell on the jaunt is found wanting and Johnson discusses the problems arising from "antiquarian credulity," the acceptance of untested opinion, the intellectual indolence which remains satisfied by rude guesswork. Hector Boethius is taken to task for his inaccuracy:

> *Lough Ness* is about twenty-four miles long, and from one mile to two miles broad. It is remarkable that *Boethius*, in his description of Scotland, gives it twelve miles of breadth. When historians or geographers exhibit false accounts of places far distant, they may be forgiven, because they can tell but what they are told; and that their accounts exceed the truth may be justly supposed, because most men exaggerate to others, if not to themselves: but *Boethius* lived at no great distance; if he never saw the lake, he must have been very incurious, and if he had seen it, his veracity yielded to very slight temptations (p. 22).

It is important to note that though Johnson holds Boethius culpable for the "fabulousness" of his Scottish history, his credulity is excusable: "Learning was then rising on the world; but ages so long accustomed to darkness, were too much dazzled with its light to see any thing distinctly . . . . The contemporaries of Boethius thought it sufficient to know what the ancients had delivered. The examination of tenets and of facts was reserved for another generation" (p. 11). As he writes later, "To count is a modern practice, the ancient method was to guess: and when numbers are guessed they are always magnified."[41] The practices of his older predecessors are associated with barbarism; Johnson prefers the methodology of the new philosophy, which not only calls many things in doubt but relies on exact observation for the establishment of truth: "It is true that of far the greater part of things, we must content ourselves with such knowledge as description may exhibit, or analogy supply; but it is true likewise, that these ideas are always incomplete, and that at least, till we have compared them with realities, we do not know them to be just" (p. 29). This insistence on accuracy and the scrupulous separation of truth from falsehood are, of course, not specifically scientific virtues; such concerns did not materialize suddenly in the Renaissance. However, science did so much to deepen and extend awareness of these matters that it is not surprising that Johnson associates the general norms with the specific intellectual tradition and casts his own position in chronological terms, with a period of intellectual immaturity and laxness preceding an era of enlightened exactitude.

Johnson's stress on precision at the expense of stylistic concerns has already been discussed. The passages in which he exhibits his attention to detail, such as the discussions of Scottish brogues and the "incommodiousness of . . . Scotch windows" are known to all readers of the *Journey*. He exhorts every writer-traveler to carry proper instruments for the measurement of heights and distances (p. 109) and approaches the accounts of others with a hearty skepticism: "We were told that [*Lough Ness*] is in some places a hundred and forty fathoms deep, a profundity scarcely credible, and which probably those that relate it have never sounded" (p. 22). "These caves were represented to us as the cabins of the first rude inhabitants, of which, however, I am by no means persuaded" (p. 54). "A single meal of a goat is a quart, and of a sheep a pint. Such at least was the account, which I could extract from those of whom I am not sure that they ever had inquired" (pp. 61–62). "The common opinion is, that peat grows again where it has been cut; which,

41. Ibid., p. 74. Cf. *Life*, III, 356: "The information which we have from modern travellers is much more authentick than what we had from ancient travellers; ancient travellers guessed; modern travellers measure."

as it seems to be chiefly a vegetable substance, is not unlikely to be true, whether known or not to those who relate it."[42]

Johnson's famous distrust of the authenticity of the Ossian poems demonstrates very clearly his demand for concrete evidence and his refusal to accept blindly a belief that had notable adherents. It illustrates the manner in which he carefully assesses the facts, possibilities, and available evidence before giving the reader his considered judgment, a process which recurs throughout the *Journey*. When the evidence attending a particular phenomenon or event is inadequate, he does not hesitate to speculate or imaginatively reconstruct, but he recognizes and admits that his tentative judgment is nothing more than educated guesswork. Like his great seventeenth-century predecessors Bacon, Browne, and Bayle, he takes pains to debunk established but unsupported beliefs: "Honesty is not greater where elegance is less" (p. 31). "We observed in travelling, that the nominal and real distances of places had very little relation to each other" (p. 45). "The cattle of *Sky* are not so small as is commonly believed" (p. 61). "It is generally supposed, that life is longer in places where there are few opportunities of luxury; but I found no instance here of extraordinary longevity" (p. 63). "I do not find it to be true, as it is reported, that to the *Second Sight* nothing is presented but phantoms of evil."[43]

Immediately preceding what is probably the most often quoted passage in the *Journey*—the lines on Marathon and Iona —Johnson states that, "Whatever withdraws us from the power of our senses; whatever makes the past, the distant, or the future predominate over the present, advances us in the dignity of thinking beings" (p. 111). Empiricism is not its own end. Johnson's method goes hand in hand with the desired results of his observations. The activity of the senses is subordinate to the reflections which arise as a result of that activity. "Far from me and from my friends, be such frigid philosophy as may conduct us indifferent and unmoved over any ground which has been dignified by wisdom, bravery, or virtue." And the reflection

42. *Journey*, p. 76. Cf. pp. 47, 72, 73–74, 82, 88, 113, 117.
43. Ibid., p. 81. Cf. pp. 48, 55–56, 77, 79, 87, 117.

which follows is only one of many to which readers of the *Journey* frequently return.

These reflections are so rich and various that one need quote only a few: "What cannot be done without some uncommon trouble or particular expedient, will not often be done at all" (p. 16). "If an epicure could remove by a wish, in quest of sensual gratifications, wherever he had supped he would breakfast in Scotland" (p. 42). "In travelling even thus almost without light thro' naked solitude, when there is a guide whose conduct may be trusted, a mind not naturally too much disposed to fear, may preserve some degree of cheerfulness; but what must be the solicitude of him who should be wandering, among the craggs and hollows, benighted, ignorant, and alone?" (p. 57). "Without intelligence man is not social, he is only gregarious, and little intelligence will there be, where all are constrained to daily labour, and every mind must wait upon the hand" (p. 102). But these reflections are firmly tied to particular observations. Johnson, the master of the generalization, is dependent upon Johnson the skeptic, the debunker, the discriminating empiricist. The passage from experience to commentary is both vigorous and dynamic. The Johnson who emerges in the *Journey* is properly described in a passage from one of the most famous entries in the *Encyclopédie*:[44]

Le *philosophe* forme ses principes sur une infinité d'observations particulières. Le peuple adopte le principe sans penser aux observations qui l'ont produit: il croit que la maxime existe, pour ainsi dire, par elle-même; mais le *philosophe* prend la maxime dès sa source; il en examine l'origine; il en connaît la propre valeur, et n'en fait que l'usage qui lui convient.

La vérité n'est pas pour le *philosophe* une maîtresse qui corrompe son imagination, et qu'il croit trouver partout; il se contente de la pouvoir démêler où il peut l'apercevoir. Il ne la confond point avec

---

44. The famous article "Philosophe," long considered to be the work of Diderot, has also been attributed to Dumarsais. See Herbert Dieckmann, *"Le Philosophe": Texts and Interpretation*, Washington University Studies, NS, Lang. and Lit., No. 18 (St. Louis, 1948). Dieckmann does not suggest a definite attribution but notes the arguments for Diderot and Dumarsais, and the possibility that the article might be the work of La Mettrie.

la vraisemblance; il prend pour vrai ce qui est vrai, pour faux ce qui
est faux, pour douteux ce qui est douteux, et pour vraisemblable ce
qui n'est que vraisemblable. Il fait plus, et c'est ici une grande perfection du *philosophe*, c'est que lorsqu'il n'a point de motif propre pour
juger, il sait demeurer indéterminé.[45]

Despite the fact that the *Journey* has been used in the past
to stress Johnson's intellectual limitations, by illustrating some
of his "prejudices," it is, in many ways, the most useful entrée
into his thought. The empirical temperament, cautious with regard to received opinion, founding its knowledge on experience,
and rising to synthesis or generalization only when possible, and
then with qualification; the dynamic struggle for knowledge
with full awareness of the difficulty of the search; the humility
and consciousness of limitations—all are apparent in these pages.
There are essential points of difference between Johnson and the
philosophes, but he shares, with the major figures of the Enlightenment, the analytical and "critical" temper which is traceable, both in his case and in that of the *philosophes*, to the
methods and successes of the new science.

Bacon promised that such success would result from "con-
junction of labors"; cooperation, diligence, and most important
of all, the guidance provided by the proper method, would
assure fulfillment of his dream. The process would be conducted
in two stages. First, the universal natural history (a work of

45. The philosopher forms his principles upon an infinity of particular
observations. Most people adopt the principle with no awareness of the
observations which have produced it. They believe that the maxim exists,
so to speak, by itself. But the philosopher takes the maxim at its source.
He examines its origin; he knows its proper value, and only makes use
of it when he feels it is appropriate.

Truth is not, for the philosopher, a mistress who corrupts his imagination, whom he believes he finds everywhere. He is content to be able
to discern it where he can perceive it. He does not confuse it with its
shadow. He considers true that which is true, false that which is false,
doubtful that which is doubtful, and probable that which is only probable.
He does more, and that is the great perfection of the philosopher, that
when he has no proper grounds for judgment, he is wise enough to remain
undecided.

generations) would be completed. Then induction would be employed on the materials of the natural history and, in a matter of only a few years, knowledge of nature would be man's. The work calls for two types of scientists. To compile the natural history no unusual ability would be necessary, but the theorists who would study the materials provided by the observers and experimenters would be men of superior ability.[46]

As was indicated earlier, Bacon's ideas were subject to considerable distortion. His stress on utilitarianism, despite the careful distinction between experiments of light and fruit, could easily be debased by men of a particular temper. Also, it was sometimes assumed, on the basis of Bacon's program, that the collected works of a large number of tinkerers would result in the acquisition of more truth than the work of a single genius, that sheer numbers were more important than qualifications.[47] Of course, some of Bacon's notions, as he articulated them, are open to searching criticism. The length of his estimated time span was impossibly idealistic; his aversion to specialization was a short-lived dream and, in that connection, his misunderstanding of William Gilbert's achievement is indeed unfortunate. Johnson's comments on these principles—the importance of co operation, the scientific time span, utilitarian applications, the qualifications necessary for the scientific student—are extremely balanced; he purges them of that degree of exaggeration upon which the polemicist usually seizes.

Johnson has little sympathy for cooperative projects, a particularly troublesome issue since it indicates Bacon's tendency to trust in organization to the neglect of the personal dimension in scientific study. Johnson, who had, of course, observed the accomplishments of Boyle and Newton, places greater faith in the work of an individual than in the concerted efforts of a group. In the *Life of Hughes* he discusses Tonson's project for a translation of Lucan's *Pharsalia* by several hands. This design, "as must often happen where the concurrence of many is neces-

46. Richard Foster Jones, *Ancients and Moderns*, 2nd ed. (St. Louis, 1961), p. 288, n. 69.
47. Mazzeo, *Renaissance and Revolution*, p. 220.

sary, fell to the ground; and the whole work was afterwards performed by Rowe."[48] Johnson determined to "confide in [himself], and no longer . . . solicit auxiliaries, which produced more incumbrance than assistance"[49] when he compiled his *Dictionary*, and felt that in England "an academy could be expected to do but little."[50] He discusses cooperative undertakings at length in *Adventurer* 45:

> The speculatist, when he has carefully observed how much may be performed by a single hand, calculates by a very easy operation the force of thousands, and goes on accumulating power till resistance vanishes before it; then rejoices in the success of his new scheme, and wonders at the folly or idleness of former ages, who have lived in want of what might so readily be procured, and suffered themselves to be debarred from happiness by obstacles which one united effort would have so easily surmounted.
>
> But this gigantic phantom of collective power vanishes at once into air and emptiness, at the first attempt to put it into action. The different apprehensions, the discordant passions, the jarring interests of men, will scarcely permit that many should unite in one undertaking (*Adventurer*, pp. 357–58).

In doubting the value of cooperative work Johnson is thinking of joint operations wherein the work is divided among several or many men. To the larger type of cooperation, that which extends over a period of time, he is, of course, wholly sympathetic, but to expect a group of learned men to join harmoniously and agree completely is folly. Discord always preponderates over unanimity and, despite the pretensions of cooperation, one man very often performs the major share of the task. Johnson is certainly not encouraging self-sufficiency. He is raising a very practical issue, one to which he brings extensive personal experience, one which involves the deepest

48. *Lives of the English Poets, by Samuel Johnson, LL.D.*, ed. G. B. Hill, 3 vols. (Oxford, 1905), II, 162. [Cited hereafter in text and notes as *Lives of the Poets.*]

49. 1825 *Works*, V (*Preface to the English Dictionary*), 43. The "auxiliaries" refer to books as well as individuals.

50. *Lives of the Poets*, I (*Life of Roscommon*), 233.

elements of human nature: "different apprehensions," "passions," conflicting interests, and ultimate isolation. This particular type of discourse, it might be added, is one in which Johnson is seldom surpassed.

To joint projects Johnson opposes the cooperation which spans centuries and geographical barriers, the slow accumulation of knowledge. Human knowledge advances at a humbling pace: "Every science was thus far advanced towards perfection, by the emulous diligence of contemporary students, and the gradual discoveries of one age improving on another" (*Rambler*, v, no. 154, 57). The natural philosopher should attempt small tasks and pursue them assiduously if great contributions are to be made: "The chief art of learning, as Locke has observed, is to attempt but little at a time. The widest excursions of the mind are made by short flights frequently repeated; the most lofty fabricks of science are formed by the continued accumulation of single propositions" (*Rambler*, iv, no. 137, 361). Johnson continually alludes to the slow process of attaining knowledge. He mentions "the scantiness of knowledge,"[51] the "Arabian proverb, that drops added to drops constitute the ocean,"[52] and "the slow advances of truth."[53] "If what appears little be universally despised, nothing greater can be attained, for all that is great was at first little, and rose to its present bulk by gradual accessions, and accumulated labours" (*Rambler*, iv, no. 83, 72). If the rate at which knowledge advances is sobering, the fact that there is so much to be done can serve as an incentive:

Electricity is the great discovery of the present age, and the great object of philosophical curiosity. It is perhaps designed by providence for the excitement of human industry, that the qualities of bodies should be discovered gradually from time to time. How many wonders may yet lie hid in every particle of matter no man can determine. The power of Electricity is sufficient to shew us that nature is far from being exhausted, and that we have yet much to

51. 1825 *Works*, v (*The Plan of an English Dictionary*), 22.
52. Ibid., p. 18.
53. *Johnson on Shakespeare*, vii, 99.

do before we shall be fully acquainted with the properties of these things which are always in our hands and before our eyes.[54]

The difficulties in attaining knowledge may be chastening, but success does occur and when Johnson describes the fruitful accumulation of knowledge he rises to great eloquence:

It is pleasing to contemplate a manufacture rising gradually from its first mean state by the successive labours of innumerable minds; to consider the first hollow trunk of an oak, in which, perhaps, the shepherd could scarce venture to cross a brook swelled with a shower, enlarged at last into a ship of war, attacking fortresses, terrifying nations, setting storms and billows at defiance, and visiting the remotest parts of the globe. And it might contribute to dispose us to a kinder regard for the labours of one another, if we were to consider from what unpromising beginnings the most useful productions of art have probably arisen. Who, when he saw the first sand or ashes, by a casual intenseness of heat melted into a metalline form, rugged with excrescences, and clouded with impurities, would have imagined, that in this shapeless lump lay concealed so many conveniencies of life, as would in time constitute a great part of the happiness of the world? Yet by some such fortuitous liquefaction was mankind taught to procure a body at once in a high degree solid and transparent, which might admit the light of the sun, and exclude the violence of the wind; which might extend the sight of the philosopher to new ranges of existence, and charm him at one time with the unbounded extent of the material creation, and at another with the endless subordination of animal life; and, what is yet of more importance, might supply the decays of nature, and succour old age with subsidiary sight. Thus was the first artificer in glass employed, though without his own knowledge or expectation. He was facilitating and prolonging the enjoyment of light, enlarging the avenues of science, and conferring the highest and most lasting pleasures; he was enabling the student to contemplate nature, and the beauty to behold herself (*Rambler*, iii, no. 9, 49–50).

Wimsatt has noted Johnson's periphrastic strategy in this passage, withholding the word "glass" until the end, as if in answer

<hr>

54. Review of R. Lovett, *The Subtil Medium proved* . . . , *Literary Magazine*, 1, v, 231–32. Attributed by Donald Greene, "Johnson's Contributions to the *Literary Magazine*," *Review of English Studies*, NS 7 (Oct. 1956), 387.

to a poetic riddle.[55] It illustrates very well Johnson's ability to modulate diction and accumulate example as he comprehends his subject in a series of diverse contexts. The relative scarcity in literature of effective and convincing paeans to technology is an indication of his skill.

In his desire for a utilitarian end for science Johnson is at one with Bacon; he constantly assesses the pragmatic value of an endeavor: "The value of a work must be estimated by its use . . . ."[56] "The only end of writing is to enable the readers better to enjoy life, or better to endure it . . . ."[57] Rasselas argues that "It is our business to consider what beings like us may perform . . . by promoting within [each man's] circle, however narrow, the happiness of others" (*Rasselas*, p. 67). Of Actuarius Johnson says:

> In the Conclusion of his Discourse upon *Urines*, he speaks with a just Severity of those that engross Truth and Science, and are displeased with any Improvements made public for the Benefit of Mankind. The Slanders of these Men, says he, are Infections, against which it would be more for the Interest of the World to find an Antidote, than against any Contagion or Disease . . . .[58]

He writes of Boerhaave that his knowledge of botany "was not of the barren Kind; for it furnish'd him with new Subjects for Chymical Operations, and new Medicines for Use,"[59] and in commenting on Lucas' *Essay on Waters* he judges the utilitarian findings more important than the chemical ones: "It is of more importance to know what diseases these waters will cure than of what ingredients they are compounded . . . ."[60] His chief remark on Payne's *Elements of Trigonometry* concerns the usefulness of mathematics: "Among the several branches [of mathematics], the excellence of Trigonometry, and its usefulness to

55. W. K. Wimsatt, *Philosophic Words* (New Haven, 1948), p. 76, n. 27. Wimsatt provides parallels (pp. 76–77) to this passage in Bacon, Boyle, Ray, Chambers, and Antonio Neri.

56. 1825 *Works*, v (*The Plan of an English Dictionary*), 3.

57. Ibid., vi (*Review of a Free Enquiry*), 66.

58. "Actuarius," *Med. Dict.*, i, sig. Ii2^r.

59. "*Botany*," ibid., sig. 10E1^r.

60. *Literary Magazine*, 1, vi, 291.

the commerce of mankind, are so apparent; that it is not un-worthy the attention of persons in the most elevated stations."[61]

Johnson's sense of duty to mankind is a constant and imposing concern. It has been argued that the major theme of his periodical essays is "contributing to society."[62] In this connection, his de-sire to see knowledge propagated is a corollary to his wish for practical applications of science, for "knowledge [is] nothing but as it is communicated . . ." (Rasselas, p. 85). He attempts to disseminate and "popularize" knowledge in his own works:

> . . . I shall not think my employment useless or ignoble, if, by my assistance, foreign nations, and distant ages, gain access to the propa-gators of knowledge, and understand the teachers of truth; if my labours afford light to the repositories of science, and add celebrity to Bacon, to Hooker, to Milton, and to Boyle.[63]

In discussing Milton's versification he attempts to avoid tech-nical terms since he is "desirous to be generally understood . . ." (Rambler, IV, no. 86, 89), and he hopes that he has made Shake-speare's meaning "accessible to many who before were frighted from perusing him, and contributed something to the publick, by diffusing innocent and rational pleasure."[64] He reserves spe-cial praise for popularizers: "To understand the works of cele-brated authors, to comprehend their systems, and retain their reasonings, is a task more than equal to common intellects; and he is by no means to be accounted useless or idle, who has stored his mind with acquired knowledge, and can detail it occasionally to others who have less leisure or weaker abilities."[65]

In a review of Stephen White's Collateral Bee-Boxes, John-son describes the "reverend author" as "a man of ingenuity,

61. Prefaces & Dedications, p. 154. Cf. the Dedication to Payne's In-troduction to Geometry, par. 3, p. 152.

62. A. T. Elder, "Thematic Patterning and Development in Johnson's Essays," Studies in Philology, 62 (July 1965), 630.

63. 1825 Works, v (Preface to the English Dictionary), 49–50.

64. Johnson on Shakespeare, VII, 103.

65. Adventurer 85, p. 413. Cf. 1825 Works, v (Considerations on the Case of Dr. Trapp's Sermons), 466 (item 26), and Adventurer 137, pp. 491–92.

candor, and, what is far more valuable, of piety; willing to communicate his knowledge for the advantage of others . . . ."[66] Communication is so important that Johnson takes scientific writers to task if they in any way hinder their readers from understanding them. Home, in his *Experiments on Bleaching*, uses too many technical terms[67] and Lucas, in his *Essay on Waters*, employs a curious orthography which diminishes the pleasure of reading his book.[68] In *Adventurer* 85 he notes Boerhaave's complaint that many writers on chemistry are useless to most readers since they presuppose more knowledge than is reasonable. They are preoccupied with their subject, and think their readers have been studying that subject as diligently as themselves (*Adventurer*, p. 414). Boerhaave's own practices constitute the ideal.

he treated that Science with an Elegance of Style not often to be found in chymical Writers, who seem generally to have affected not only a barbarous, but unintelligible Phrase, and, like the *Pythagoreans* of old, to have wrapt up their Secrets in Symbols, and enigmatical Expressions, either because they believed, that Mankind would reverence most what they least understood, or because they wrote not from Benevolence, but Vanity, and were desirous to be praised for their Knowledge, though they could not prevail upon themselves to communicate it.[69]

66. *Literary Magazine*, 1, i, 27. Attributed by Greene, "Johnson's Contributions to the *Literary Magazine*," p. 377. Cf. the Proposals for James's *Medicinal Dictionary* (*Prefaces & Dedications*, p. 69): "It is doubtless of Importance to the Happiness of Mankind, that whatever is generally useful should be generally known; and he therefore that *diffuses* Science, may with Justice claim, among the Benefactors to the Public, the next Rank to him that *improves* it." See also *Life*, 1, 550.

67. *Literary Magazine*, 1, iii, 136.

68. Ibid., iv, 167.

69. "Boerhaave," *Med. Dict.*, 1, sig. 9U2ᵛ. Cf. Thomas Sprat, *History of the Royal Society*, ed. Jackson I. Cope and Harold Whitmore Jones (St. Louis, 1966), p. 5: "It was the custom of their [the East's] Wise men, to wrap up their Observations on Nature, and the Manners of Men, in the dark Shadows of *Hieroglyphicks*; and to conceal them, as sacred *Mysteries*, from the apprehensions of the vulgar. This was a sure way to beget a Reverence in the Peoples Hearts towards *themselves*: but not to advance the true Philosophy of *Nature*."

In light of the Royal Society's attempts at stylistic reform, it is important to note that Johnson does not call for simplicity and homogeneity, but rather clarity, grace, and elegance.[70]

In assessing the qualifications necessary for scientific study, Johnson recognizes a spectrum of abilities. He would agree with Bacon that the industrious man of average talent can "do" science, but he would hasten to point out that some men are able to do it in much more important ways. As Imlac notes, "great works are performed, not by strength, but perseverance . . ." (*Rasselas*, p. 37). There are men who can grasp complex problems and offer solutions without the laborious research that would usually be necessary, but knowledge is primarily advanced by the diligent. The precocious, dazzling student, as Ascham had observed, is likely to disappoint expectations, while the work of the tenacious though less gifted man may in the long run prove most fruitful:

If we except those gigantick and stupendous intelligences who are said to grasp a system by intuition, and bound forward from one series of conclusions to another, without regular steps through intermediate propositions, the most successful students make their advances in knowledge by short flights between each of which the mind may lie at rest.[71]

70. The best introduction to the relations between science, language, and the Royal Society is the series of articles by R. F. Jones reprinted in *The Seventeenth Century: Studies in the History of English Thought and Literature from Bacon to Pope* (Stanford, 1951): "Science and English Prose Style in the Third Quarter of the Seventeenth Century," pp. 75–110; "The Attack on Pulpit Eloquence in the Restoration: An Episode in the Development of the Neo-Classical Standard for Prose," pp. 111–42; "Science and Language in England of the Mid-Seventeenth Century," pp. 143–60. It goes without saying that Johnson is not in sympathy with movements to reduce language to its simplest possible terms, to substitute symbols for words, to achieve absolute equivalence between a word and a thing. See W. K. Wimsatt, Jr., *The Prose Style of Samuel Johnson* (New Haven, 1941), pp. 99–100.

71. *Rambler*, iv, no. 108, 212–13. Cf. ibid., no. 111, p. 228: "Even those who are less inclined to form general conclusions, from instances which by their own nature must be rare, have yet been inclined to prognosticate no suitable progress from the first sallies of rapid wits; but have ob-

Since genius is not the primary qualification, most men are capable of pursuing scientific studies:

> A man that has formed this habit of turning every new object to his entertainment, finds in the productions of nature an inexhaustible stock of materials upon which he can employ himself . . . . He has always a certain prospect of discovering new reasons for adoring the sovereign author of the universe, and probable hopes of making some discovery of benefit to others, or of profit to himself. There is no doubt but many vegetables and animals have qualities that might be of great use, to the knowledge of which there is not required much force of penetration, or fatigue of study, but only frequent experiments, and close attention (*Rambler*, III, no. 5, 28–29).

Nevertheless, the power of every man is small: "Providence has given no man ability to do much, that something might be left for every man to do. The business of life is carried on by a general co-operation; in which the part of any single man can be no more distinguished, than the effect of a particular drop when the meadows are floated by a summer shower: yet every drop increases the inundation, and every hand adds to the happiness or misery of mankind" (*Adventurer* 137, p. 489). Everyone can, and should, add something but "to add much can indeed be the lot of few . . ." (*Rambler*, IV, no. 129, 325). The frailty of man necessitates our admiration of each of his successes, no matter how small, but our highest praise must be reserved for the man who contributes works of greatness:

> To mean understandings, it is sufficient honour to be numbered amongst the lowest labourers of learning; but different abilities must find different tasks. To hew stone, would have been unworthy of Palladio; and to have rambled in search of shells and flowers, had but ill-suited with the capacity of Newton (*Rambler*, IV, no. 83, 76).

Johnson's judgment of the nature of the relationship between science and religion, and its Baconian precedent, is the subject of the fifth chapter, but we may conclude here that Johnson

---

served, that after a short effort they either loiter or faint, and suffer themselves to be surpassed by the even and regular perseverance of slower understandings."

takes up the essential issues of Bacon's program and is in general agreement with them. There are some points of departure, but these are largely attributable to the alterations in the scientific landscape which had occurred between Bacon's time and Johnson's. (The Royal Society, for example, had recognized that Bacon's aversion to specialization was shortsighted and had set up committees to direct the various activities of the Society in their several branches. The stress of genius and its accomplishments was virtually de rigueur after Newton; it had become clear that the production of a work like the *Principia* represented a more feasible ideal than the attainment of a universal natural history through the cooperative labor of men of average ability.) Johnson's admiration of Bacon is well known: he quotes him constantly, and once contemplated the preparation of a biography and edition, at least of the English works.[72] If indeed he did not know Bacon's work until the compilation of the *Dictionary*, his discovery of Bacon, if it may be termed such, must have been more on the order of the recognition of a kindred spirit than a sudden stumbling upon fresh concepts, for one can be Baconian before reading Bacon. Johnson's interest in Bacon is normally associated with the *Dictionary*—the highly favorable notice Bacon receives in the Preface and the considerable number of instances in which he is cited in the text—but the norms to which Johnson subscribes in the medical biographies are in the tradition of English science and its great progenitor. A slackening of English interest in the dialectic of seventeenth-century scientific controversy focuses greater attention on those mid- and late-eighteenth-century figures who continue the discussion. Hence R. F. Jones's interest in Young's *Conjectures on Original Composition*.[73] I would suggest that the clash of Ancient and

72. Johnson's interest in and affection for Bacon is omnipresent in his work. For a brief treatment, see W. B. C. Watkins, *Johnson and English Poetry before 1660* (Princeton, 1936), pp. 59–60.

73. See his "Science and Criticism in the Neo-Classical Age of English Literature," in *The Seventeenth Century*, pp. 65–71. The influence of Bacon on the eighteenth century has not yet been delineated satisfactorily. Rexmond C. Cochrane's "Francis Bacon in Early Eighteenth-Century English Literature," *Philological Quarterly*, 37 (Jan. 1958), 58–79, should be approached with great caution.

Modern and the influence of Baconian ideology, whether implicit or explicit, is also evident in, among other works, Johnson's *Life of Boerhaave* and *Journey to the Western Islands of Scotland*. His statements are often scattered but when they are assembled they form a coherent whole; any survey of the English tradition of scientific ideology must take them into account.

CHAPTER IV

The Satiric Reaction

In his discussion of the age's "*Wits* and *Railleurs*," Sprat records a genuine fear of their assaults on the new science; searching criticism does not enjoy the power which they wield, that of laughter and ridicule:

> I acknowledge that we ought to have a great dread of their power: I confess I believe that *New Philosophy* need not . . . fear the pale, or the melancholy, as much as the humorous, and the merry: For they perhaps by making it ridiculous, becaus it is *new*, and becaus they themselves are unwilling to take pains about it, may do it more injury than all the Arguments of our severe and frowning and dogmatical *Adversaries*.[1]

The Royal Society's philosophy was new, but scientific satire most certainly was not. Jonson's Subtle had dealt in dreams and illusions; Chaucer's Canon's Yeoman recounts the fact that his lord "hath ymaad us spenden muchel good,/For sorwe of which

1. Thomas Sprat, *History of the Royal Society*, ed. Jackson I. Cope and Harold Whitmore Jones (St. Louis, 1966), p. 417.

almoost we wexen wood,/But that good hope crepeth in oure
herte,/Supposynge evere, though we sore smerte,/To be re-
leeved by hym afterward."[2] Quack physicians, purveyors of
panaceas, and fanciful schemers have never been in short supply.
Swift's view of science as a major point of contention in the
conflict between Ancient and Modern was hardly innovative. In
The Clouds, for example, Aristophanes contrasts the Athens of
the men of Marathon with the Athens of sophistical education.
"Wrong Logic," the personification of the sophistical method,
demonstrates how "old established rules and laws" may be
easily contradicted. His tool is the manipulation of words; he
deals in chop-logic and casuistry rather than substance. Pheidippides is justly surprised at the fact that the chief bit of learning
his father Strepsiades has acquired at Socrates' Phrontisterion
(thinking-house) is that a female fowl is a "fowless"—one of
the "mighty secrets" in the eyes of Pheidippides. But knowledge
of the "Laws of Nature" is also casuistry—words, not substance—
so that Strepsiades haughtily challenges his creditor Amynias'
ignorance of theories concerning the relation between the sun
and rainfall. (Appropriately, the clouds are the sophists' patrons.) At the Phrontisterion the students study fleas and gnats,
fix their eyes on the ground, "diving deep into the deepest
secrets," and with their rumps turned toward the sky take "private lessons on the stars." Their projects are on the one hand
incredibly trivial, on the other ridiculously grandiose. Their
great mentor, Socrates, walks on air, contemplates the sun,
searches out celestial matters, and infuses his "subtle spirit with
the kindred air," taking his grand place in the cosmic scheme
betwixt heaven and earth—suspended from the ceiling in a bas-
ket. We are only a short distance from Lagado.

Nevertheless, the number of attacks on science had multiplied
in the days following the chartering of the Royal Society. In
his ode prefaced to Sprat's history, Cowley shows little patience
with such detractors: "Mischief and tru Dishonour fall on those/
Who would to laughter or to scorn expose/So Virtuous and so

2. *The Canon's Yeoman's Tale*, ll. 868–72, *The Works of Geoffrey
Chaucer*, ed. F. N. Robinson, 2nd ed. (Boston, 1957), p. 217.

Noble a Design,/So Human for its Use, for Knowledge so Divine" (ll. 151–54). His claim that philosophy is both virtuous and noble, both useful and divine, is particularly appropriate, for the major charges of the satirists were to focus on science's bathetic uselessness and self-sufficient pretentiousness. When Gulliver explained "our several systems of *natural philosophy*" to his Houyhnhnm master, "he would laugh that a creature pretending to *reason* should value itself upon the knowledge of other people's conjectures, and in things where that knowledge, if it were certain, could be of no use" (iv, ch. viii).

Perhaps the most memorable representative of useless speculation is Shadwell's Sir Nicholas Gimcrack. His wife describes the virtuoso's aquatic pursuits: "He has a frog in a bowl of water, tied with a packthread by the loins, which packthread Sir Nicholas holds in his teeth, lying upon his belly on a table; and as the frog strikes, he strikes; and his swimming master stands by to tell him when he does well or ill."[3] Gimcrack, whom Gresham College out of envy refused, contents himself "with the speculative part of swimming; I care not for the practic. I seldom bring anything to use; 'tis not my way. Knowledge is my ultimate end."[4] The danger of pride is as great as the problem of triviality. Pope can encourage potential students to forget mites and flies, and seek a larger perspective, but he, like Swift, assaults pride as well as dulness; he is not alone in suggesting that if science must be studied its everpresent guide should be Modesty. Neither the tinkering virtuoso nor the haughty projector—Swift often uses the terms interchangeably[5]—will bring forth works of fruit, and both are neglecting far more important human concerns. The point at issue, in the satirist's eyes, is that of a fractured value system, the love of the petty or the impossible which warps

3. *The Virtuoso*, ed. Marjorie Hope Nicolson and David Stuart Rodes (Lincoln, 1966), ii, i, 295–99.

4. Ibid., ii, 84–86.

5. Dorothy Stimson, *Scientists and Amateurs* (New York, 1948), p. 131; Miriam Kosh Starkman, *Swift's Satire on Learning in "A Tale of a Tub"* (Princeton, 1950), p. 72. On the nature of the virtuoso see Walter E. Houghton, Jr., "The English Virtuoso in the Seventeenth Century," *Journal of the History of Ideas*, 3 (Jan., April 1942), 51–73, 190–219.

personality and obscures or destroys the received values which the humanist, to use Fussell's term, reveres. Shenstone raises the issue in a frivolous context:

> Let FLAVIA'S eyes more deeply warm,
>   Nor thus your hearts determine,
> To slight dame nature's fairest form,
>   And sigh for nature's vermin.[6]

Like his literary betters, Stephen Duck attempts to readjust the perspective:

> DEAR Madam, did you never gaze,
> Thro' Optic-glass, on rotten *Cheese*?
> There, Madam, did you ne'er perceive
> A Crowd of dwarfish Creatures live?
> The little Things, elate with Pride,
> Strut to and fro, from Side to Side:
> In tiny Pomp, and pertly vain,
> Lords of their pleasing Orb, they reign;
> And, fill'd with harden'd Curds and Cream,
> Think the whole Dairy made for *them*.
> So Men, conceited Lords of all,
> Walk proudly o'er this pendent Ball,
> Fond of their little Spot below,
> Nor greater Beings care to know;
> But think, those *Worlds*, which deck the Skies,
> Were only form'd to please *their* Eyes.[7]

As Marjorie Nicolson pointed out over thirty-five years ago, the possibility of viewing man and man's place in multiple perspective, which was increased immeasurably by telescopic and microscopic developments, is not only a Popean commonplace but one of the mightiest lessons of Gulliver's account of his voyages.[8]

6. Shenstone, "To the Virtuosos," ll. 37–40, in *The Works in Verse and Prose*, 3rd ed. (London, 1768), I, 206.

7. Duck, "On Mites, To a Lady," *Poems on Several Occasions* (London, 1736), pp. 160–61.

8. "The Microscope and English Imagination," *Science and Imagination* (Ithaca, 1956), pp. 193–99 (the monograph first appeared in Smith College Studies in Modern Languages in 1935).

Many of the satires, of course, are both shortsighted and inaccurate. At Lagado there is an astronomer "who had undertaken to place a sundial upon the great weathercock on the townhouse, by adjusting the annual and diurnal motions of the earth and sun, so as to answer and coincide with all accidental turnings by the wind." Sun-clocks had been invented in both France and England, and in 1663 Sir Christopher Wren had fashioned an ingenious wind recorder by annexing a clock to a weathercock, attaching a pencil to the clock and a paper on a rundle moved by the weathercock.[9] The astronomer's project was hardly an impossible one. Bonamy Dobrée comments that "We do . . . extract sunshine from cucumbers, though we put it into globules and call it Vitamin C,"[10] and the efficacy of satirizing proposed moon voyages has been dramatically called into question, but the application of hindsight in these matters is not altogether fair and I am more concerned with other forms of "inaccuracy."

In *The London Spy*, for example, Ned Ward visits "*Wiseacre's Hall*, more commonly call'd *Gresham-College*," where he encounters, along with an "*Elaboratory-keeper*," "a *Peripatetick* walking, ruminating, as I suppose, upon his *Entities, Essences*, and *occult Qualities*, or else upon the *Philosopher's Stone*; looking as if he very much wanted it . . . ."[11] Bacon's view of the Aristotelians and the alchemists could hardly be considered affectionate; Sprat takes up the matter of Aristotelian "tyranny" and, in Swiftian fashion, brands the alchemists "enthusiasts." The Society was certainly open to attack, but Ward's is misplaced, and there are some satires with which the Society's Fellows would be in total sympathy. The absurd attachment to the Ancients depicted in the Scriblerus *Memoirs*, probably the work of Arbuthnot, would hardly be congenial to the spirit of the Society, and Martinus, "the youthful Virtuoso," would not be

9. Marjorie Nicolson and Nora M. Mohler, "The Scientific Background of Swift's *Voyage to Laputa*," ibid., pp. 139–40.

10. *English Literature in the Early Eighteenth Century, 1700–1740* (New York, 1959), p. 455.

11. *The London Spy Compleat, in Eighteen Parts*, 4th ed. (London, 1753), p. 54.

approved by the English scientists.[12] The editors of *Three Hours after Marriage* note that the portrait of Fossile—the Woodward figure—is an attack on bad science alone. Fossile is "undiscriminating, old-fashioned (many of his medical theories are in fact Hermetic), snobbish, ill-informed . . . . He cares for no mechanical science but only for mythical monsters . . . ."[13]

To be sure, however, discriminating attacks, with the precise recognition of specific satiric targets and the honest search for a balanced view, were seldom achieved, and the cumulative force of even inaccurate onslaughts could outweigh any momentary defense. In the long run, most individual charges or assaults on particular philosophers could be answered, point by point, with countercharges and counter examples. The rejection of science, as Blake fully realized, must be a total, not a piecemeal procedure. Nevertheless, the very bulk of satiric commentary represented a formidable challenge, which Johnson, for one, did not hesitate to take up. His recorded reactions to the satirists of science are consistently and systematically unfavorable.

Johnson characteristically defends all intellectual undertakings. Since the attainment of knowledge is exceedingly difficult, and man is eternally beset by indolence and external obstructions, anyone who attempts a project or course of study deserves encouragement. Johnson comments frequently on those who ridicule or object to intellectual endeavors: "To object, is always easy, and, it has been well observed by a late writer, that 'the hand which cannot build a hovel, may demolish a temple.'" [14] "Men who want Merit themselves, are only fond

12. For Cornelius' "superstitious veneration for the Ancients," see *Memoirs of . . . Martinus Scriblerus*, ed. Charles Kerby-Miller (New Haven, 1950), p. 125. Scriblerus "may well be called *The Philosopher of Ultimate Causes*, since by a Sagacity peculiar to himself, he hath discover'd Effects in their very Cause; and without the trivial helps of Experiments, or Observations, hath been the Inventor of most of the modern Systems and Hypotheses" (p. 166).

13. *Three Hours after Marriage*, ed. Richard Morton and William M. Peterson, Lake Erie College Studies, 1 (Painesville, Ohio, 1961), pp. iv–v.

14. 1825 *Works*, VI (*Review of a Free Enquiry*), 76.

of stripping others of theirs . . . ."[15] "Men are generally idle, and ready to satisfy themselves, and intimidate the Industry of others, by calling that impossible which is only difficult."[16] In *Adventurer* 99 he defines a project as something "attempted without previous certainty of success." It is a good thing which all too often finds detractors. He speaks of "narrow minds" and their reactions:

[A project] may, therefore, expose its author to censure and contempt; and if the liberty of laughing be once indulged, every man will laugh at what he does not understand, every project will be considered as madness, and every great or new design will be censured as a project. Men, unaccustomed to reason and researches, think every enterprise impracticable, which is extended beyond common effects, or comprises many intermediate operations.[17]

Boswell says that "Johnson seldom encouraged general censure of any profession..." (*Life*, IV, 313), and Mrs. Thrale notes his aversion to "general satire."[18] Considering his interest in and knowledge of the new philosophy, it comes as little surprise that he criticizes those who attack science or scientists. Of Butler he writes:

Some verses, in the last collection [the 1759 *Genuine Remains*], shew him to have been among those who ridiculed the institution of the Royal Society, of which the enemies were for some time very numerous and very acrimonious; for what reason it is hard to conceive, since the philosophers professed not to advance doctrines but to produce facts; and the most zealous enemy of innovation must admit the gradual progress of experience, however he may oppose hypothetical temerity.[19]

He strongly objects to the Scriblerian attack on John Woodward in *Three Hours after Marriage*:

15. "Botany," *Med. Dict.*, I, sig. 10E1r.
16. "Boerhaave," ibid., sig. 9U2v.
17. *Adventurer*, p. 434. Cf. *Prefaces & Dedications* (Introduction to *The World Displayed*), pp. 236–37, where Johnson discusses Columbus' treatment as "a fanciful and rash projector," and his eventual vindication.
18. Hester Lynch Piozzi, *Anecdotes of the Late Samuel Johnson, LL.D.*, ed. G. B. Hill in *Johnsonian Miscellanies* (Oxford, 1897), I, 223, 327.
19. *Lives of the Poets*, I (*Life of Butler*), 208–9.

Not long afterwards (1717) he [Gay] endeavoured to entertain the town with *Three Hours after Marriage*, a comedy written, as there is sufficient reason for believing, by the joint assistance of Pope and Arbuthnot. One purpose of it was to bring into contempt Dr. Woodward, the Fossilist, a man not really or justly contemptible. It had the fate which such outrages deserve: the scene in which Woodward was directly and apparently ridiculed, by the introduction of a mummy and a crocodile, disgusted the audience, and the performance was driven off the stage with general condemnation.[20]

He criticizes portions of the Scriblerus *Memoirs* that attack certain scientific aberrations and further satirize Woodward: "The follies which the writer ridicules are so little practised that they are not known; nor can the satire be understood but by the learned: he raises phantoms of absurdity, and then drives them away."[21]

Johnson's unfavorable estimate of Swift is well known. Among other things, "he always hated and censured Swift for his unprovoked bitterness against the professors of medicine . . . ."[22] He considered both Swift and Pope narrow-minded,[23] and the Scriblerus Club, which spawned so many satires on science, a haughty coterie:

From the letters that pass between him [Swift] and Pope it might be inferred that they, with Arbuthnot and Gay, had engrossed all the understanding and virtue of mankind, that their merits filled the world; or that there was no hope of more. They shew the age involved in darkness, and shade the picture with sullen emulation.[24]

---

20. Ibid., II (*Life of Gay*), 271–72. It is likely that Johnson was not particularly conversant with Woodward's personality. He was rude, arrogant, and combative. This, combined with the fact that despite his genuine achievements in geology and paleontology he was prone to fantastic theorizing, marked him as an eminent and appropriate butt for satire. See Sir Henry Lyons, *The Royal Society, 1660–1940* (Cambridge, 1944), pp. 99–100; Kerby-Miller, *Memoirs of . . . Martinus Scriblerus,* p. 204.

21. *Lives of the Poets*, III (*Life of Pope*), 182.

22. *Anecdotes*, p. 223.

23. *Lives of the Poets*, III (*Life of Pope*), 212. Cf. *Anecdotes*, p. 184.

24. *Lives of the Poets*, III (*Life of Swift*), pp. 61–62.

It is sometimes suggested that Swift's antipathy to Dryden is a result of Dryden's Modernism and affiliation with the Royal Society. It is quite likely that Johnson's view of Swift results from the latter's elaborate assaults on science, and the goals, norms, and attitudes of the Moderns. To Dryden, on the other hand, Johnson is especially receptive. The basis for Johnson's predictable response to a Swiftian work may well be traceable to this fundamental difference in intellectual temper or orientation.

Johnson's implicit history of eighteenth-century literature, and the value judgments on which it is based, differ markedly from our own. He judged Arbuthnot "the first man" among the writers of Queen Anne's reign.[25] Arbuthnot was royal physician, a scientist, mathematician, antiquarian, member of the Royal Society and Fellow of the College of Surgeons. The breadth of his knowledge and wit, as well as his skill in medicine, greatly appealed to Johnson. I think it is important that he singles out Arbuthnot for such praise. He does not say so explicitly, but he is perhaps more willing to forgive Arbuthnot for his part in the Scriblerian satires on the new science because Arbuthnot was, obviously, very sympathetic to "philosophy." He did not "want Merit," was not "generally idle," and was not "unaccustomed to reason and researches." He did not reject Modernism wholesale. I am not attributing these dire characteristics to Swift, Pope, and Gay—nor does Johnson, explicitly— but only indicating that Arbuthnot is not a member of that somewhat amorphous mob of "Wits and Railleurs" whose reactions to science are colored by personal situation and not informed by first-hand experience with the matters in question. Professionals may criticize one another; railleurs would do well to keep their cavils to themselves, for though they may provide momentary humor their comments, in Johnson's judgment, are in no way constructive.

William R. Keast writes that Johnson, "unlike Pope and Swift, never allowed himself to be betrayed into unqualified attack on

25. *Life*, I, 425. Cf. the praise of Arbuthnot in *Lives of the Poets*, III (*Life of Pope*), 177.

scholarship or antiquarian research."[26] His own practices are consistent with his unsympathetic attitude toward the attacks of the satirists. He encourages the most minute forms of research and attempts, like Bacon, to instill confidence in the potential student or planner of projects. He admires Richard Savage's self-confidence: "He always preserved a steady confidence in his own capacity, and believed nothing above his reach which he should at any time earnestly endeavor to attain."[27] The virtuoso and projector should push on, despite the derision of their detractors: "To have attempted much is always laudable . . . ."[28] "Nothing . . . will ever be attempted, if all possible objections must be first overcome" (*Rasselas*, p. 15). Resolution is all important: "Few things are impossible to diligence and skill" (*Rasselas*, p. 35).

"I wish well to all useful undertakings,"[29] Johnson writes, and nearly anything, properly viewed, can be considered useful. He defends William Payne's *Game of Draughts* in his Dedication of that work to the Earl of Rochford:

> Triflers may find or make any Thing a Trifle; but since it is the great Characteristic of a wise Man to see Events in their Causes, to obviate Consequences, and ascertain Contingencies, your Lordship will think nothing a Trifle by which the Mind is inured to Caution, Foresight, and Circumspection.[30]

Even though it does not teach, the apparently useless activity or study may inspire; it can show the extent of human power and industry. Thus he defends such men as the performer Johnson, who was able to ride three horses simultaneously: "Every thing that enlarges the sphere of human powers, that shows man

26. "Johnson and Intellectual History," in *New Light on Dr. Johnson,* ed. Frederick W. Hilles (New Haven, 1959), p. 252.

27. *Lives of the Poets,* ii (*Life of Savage*), 403.

28. 1825 *Works,* v (*Preface to the English Dictionary*), 42.

29. *Letters,* i, no. 107 (9 April 1757 to Charles O'Conor), 101–2. For examples of Johnson's help and encouragement, see W. B. C. Watkins, *Johnson and English Poetry before 1660* (Princeton, 1936), pp. 12–13; Keast, "Johnson and Intellectual History," p. 251.

30. *Prefaces & Dedications,* p. 150.

he can do what he thought he could not do, is valuable."[31] Sir Thomas Browne's "Of Garlands and Coronary or Garland-Plants," which Tenison included in his *Certain Miscellany Tracts* (1683), deals with "a subject merely of learned curiosity," but Johnson points out that it may provide the pleasure of reflecting on "the industry with which studious men have endeavoured to recover [ancient customs]."[32] As will be seen, Johnson's justification of apparently trifling scientific activities is an elaborate one, but his reaction to most intellectual pursuits of the sort which could easily serve as worthy material for the satirist—ostensibly mundane amusements or seemingly impossible schemes—is generally quite favorable. This sympathy is extended to the scientist and his lesser colleagues, the virtuoso and projector.

When Boswell feared that his journal contained too many minor incidents, Johnson replied, "There is nothing, Sir, too little for so little a creature as man" (*Life*, I, 433). "All knowledge," he observes, "is of itself of some value. There is nothing so minute or inconsiderable, that I would not rather know it than not" (*Life*, II, 357). In this connection he writes to Susannah Thrale:

> With Mr Herschil [*sic*] it will certainly be very right to cultivate an acquaintance, for he can show you in the sky what no man before him has ever seen, by some wonderful improvements which he has made in the telescope. What he has to show is indeed a long way off, and perhaps concerns us but little, but all truth is valuable and all knowledge is pleasing in its first effects, and may be subsequently useful . . . . Take therefore all opportunities of learning that offer themselves, however remote the matter may be from common life or common conversation. Look in Herschel's telescope; go into a chymist's laboratory; if you see a manufacturer at work, remark his operations.[33]

The mark of intellectual eminence is, in fact, the familiarity with lowly as well as lofty matters. "The true strong and sound mind is the mind that can embrace equally great things and

31. *Life*, III, 231. Cf. *Life*, I, 399.
32. 1825 *Works*, VI (*Life of Browne*), 487.
33. *Letters*, III, no. 944 (25 March 1784), 144.

small" (*Life*, III, 334). Johnson's ideal is also a self-description. He himself follows the advice which he gives Susannah Thrale. After a particularly impressive interchange in which Johnson discussed several topics, among them the gathering of bones by the poor, the reason for which he explained at great length, Boswell notes that he has recorded the incidentals of the conversation, "which some may think trifling, in order to shew clearly how this great man, whose mind could grasp such large and extensive subjects, as he has shewn in his literary labours, was yet well-informed in the common affairs of life, and loved to illustrate them" (*Life*, IV, 206). It is common now to encounter depreciations of Boswell, and many of his shortcomings are serious ones, but in this case his statement, Johnson's published pronouncements, and his easily demonstrated practices all parallel one another. Johnson's considered judgment of the value of "petty" enterprises and his notion of intellectual strength and sophistication are informed by first-hand experience in the vicissitudes brought about by penury, as well as the difficulties and obstructions facing the editor or lexicographer who attempts bold projects. He possesses the experiential knowledge he recommends; Boswell often recounts it in detail and a large portion of his genius involves his realization of the veracity of the Johnsonian notion that for the biographer seeking faithfully to depict character and personality, the importance of quotidian vagaries often outweighs that of the dramatic event. Orange peels are as important as LL.D.'s.

Johnson's openness and driving curiosity, his encouragement of others to perform all that is within their power, his sympathy for extremely diverse intellectual undertakings, and his defense of minor (and major) figures from the attacks of their critics, are central elements of his cast of mind. We can recall Raleigh's remark: "The greatness of Johnson is seen in the generosity of his temper. An intellect may be strong and active; it is only a temper that is great."[34] Johnson welcomes the intellectual advances both in science and scholarship; his skepticism, empiricism, and dislike of undue respect for tradition and authority,

34. Walter Raleigh, *Six Essays on Johnson* (Oxford, 1910), p. 173.

place him on the side of the Moderns, against his satiric predecessors, in the Ancients-Moderns controversy. Of the *Tale of
a Tub* he writes:

> The digressions relating to Wotton and Bentley must be confessed
> to discover want of knowledge or want of integrity; he did not un-
> derstand the two controversies, or he willingly misrepresented them.
> But Wit can stand its ground against Truth only a little while. The
> honours due to learning have been justly distributed by the decision
> of posterity.[35]

William King, a minor satirist of science, is similarly criticized
for entering the fray: "In 1697 he mingled in the controversy
between Boyle and Bentley; and was one of those who tried
what Wit could perform in opposition to Learning, on a question which Learning only could decide."[36] Of the *Dunciad* Johnson remarks that "the satire which brought Theobald and Moore
into contempt, dropped impotent from Bentley, like the javelin
of Priam."[37] Bentley, the most important Modern party to the
controversy, was associated with the Royal Society in several
ways. He had been chosen to deliver the first of the Boyle
lectures in 1692, and later established a biological laboratory at
Cambridge. It has been said that the scientific spirit of his scholarship reflects that of the experimental philosophers[38]—important, though partial grounds for Swift's attacking him.

Johnson "said he thought very highly of Bentley; that no man
now went so far in the kinds of learning that he cultivated; that
the many attacks of him were owing to envy and to a desire of
being known by being in competition with such a man . . ."
(*Tour*, p. 140). He praises Bentley constantly[39] and cites him
as a major example in one of his most important defenses of
modern attainments:

35. *Lives of the Poets*, III (*Life of Swift*), 11.
36. Ibid., II (*Life of King*), 27.
37. Ibid., III (*Life of Pope*), 241–42.
38. Richard Foster Jones, "The Background of *The Battle of the
Books*," *Washington University Studies*, 7, No. 2 (April 1920), 157.
39. See, for example, *Tour*, p. 307; *Life*, I, 71; II, 444; IV, 21; IV, 23–24;
Johnson on Shakespeare, ed. Arthur Sherbo, 2 vols. (New Haven, 1968),
VII, 109–110.

Men in ancient times dared to stand forth with a degree of ignorance with which nobody would dare now to stand forth. I am always angry when I hear ancient times praised at the expence of modern times. There is now a great deal more learning in the world than there was formerly; for it is universally diffused. You have, perhaps, no man who knows as much Greek and Latin as Bentley; no man who knows as much mathematicks as Newton: but you have many more men who know Greek and Latin, and who know mathematicks (*Life*, IV, 217).

"Never prone to inveigh against the present times" (*Life*, III, 3), Johnson points out the industry of modern writers and the accuracy of their methodology, preferring modern practices to the credulity which verges on barbarism. He aligns himself with the seventeenth-century scientists and ideologists in deploring the fruitless stress of rhetoric at the expense of veracity: "The first race of scholars, in the fifteenth century, and some time after, were, for the most part, learning to speak, rather than to think, and were therefore more studious of elegance than of truth."[40] He says of the barbarian invasion of the Roman empire that "their true numbers were never known. Those who were conquered by them are their historians . . ." (*Journey*, p. 73). In their attempt to achieve certainty by directing their studies with a rigorous methodology, Johnson is wholly in sympathy with the Moderns—who count rather than guess—and responds to their critics not so much in outrage as surprise. The ambitions of modern scientists and scholars are laudable; considering the difficulties attending human studies as well as the limitations besetting human nature, the withholding of encouragement or witty indulgence in attack are incomprehensible. As Carson S. Duncan wrote, Johnson "accepts the new science as a matter of course . . . ."[41]

40. *Journey*, p. 11. Cf. 1825 *Works*, v (*Preface to the English Dictionary*), 27: "I am not yet so lost in lexicography, as to forget that *words are the daughters of earth, and that things are the sons of heaven.* Language is only the instrument of science, and words are but the signs of ideas . . . ."

41. *The New Science and English Literature in the Classical Period* (Menasha, Wis., 1913), p. 176.

However, Duncan's statement has either been disputed or gone unnoticed, for it has by some been assumed that Johnson's own practices—his treatment of virtuosi and projectors, particularly in the periodical essays—associate him with the satirists of science rather than with its encomiasts. In the *Rambler*, *Idler*, and *Adventurer* essays he creates his own quasi-scientists and comments on them and their pursuits at length. Because of the number of scientific amateurs at work in the eighteenth century, the common intellectual currency which scientific notions constituted, and Johnson's use of the periodical essay form for social commentary and criticism, we can expect a certain degree of humor in the sketches of lesser students of science and collectors of antiquities. He does employ his wit and enjoys the tracing of "virtuosity" in its aberrational extremes, but the total view of the scientific hangers-on which emerges from Johnson's essays is favorable. When his comments are particularly sharp, when his rhetoric belies his basic sympathies, his tendency is to back down, qualify his statements, and set matters in proper perspective. At times he may appear to be less than enthusiastic, but considering his remarks within the context of the essays in question as well as the literary form in which he is working, and the general praise and encouragement which counterbalance the ostensible criticism, his treatment is strikingly sympathetic. His inclination is to explain and justify, not flatly to censure.

Some of the seemingly unflattering remarks on virtuosi and projectors can be dispensed with at the outset. In *Idler* 6 we learn that there is a kind of virtuoso underground. Besides the conscientious collectors there are thieves who steal curiosities and only stoop to purchase when all similar specimens are destroyed. "Tim Ranger" (*Idler* 64) tells the Idler that envious collectors have attempted to destroy his character because of his success in obtaining rarities, and Johnson warns his readers (*Idler* 14) that the man who does not withdraw from society may expect to have some of his precious time frittered away in attending to the vain expectations of projectors. These brief comments are found in more generalized discussions and appear as asides or illustrations. When Johnson devotes full attention to the activities of the minor "philosophers," the advantages and disadvantages of their pursuits are carefully assessed. Finally, he decides in favor of "projection" or "virtuosity."

The genre of the Character is particularly congenial to Johnson's talents. Miniaturized sketches representing individual approaches to experience that may easily be generalized recur throughout his work, not only in the essays, but elsewhere, for example in the satiric portraits of *London* and *The Vanity of Human Wishes*. Chesterfield as patron, Levet as humble physician, Stephen Gray as pious scientist—all are, in a sense, Characters. It could be argued that the mad astronomer, hapless stoic, and even the major personages of *Rasselas* are, to an important extent, closely related to this tradition. In this regard, rigid generic distinctions are easily blurred. However, we are particularly concerned with the gallery of quasi-"philosophers" in the periodical essays. In each case, the portraits imply no fundamental criticism of the virtuosi's or projectors' activities, since what might be construed as important charges are mitigated by context or qualifying statement. Johnson is not reticent to criticize, but the failings and affectations which are depicted are, in the largest sense, human ones, not specifically and exclusively sins or absurdities which may be attributed only to the representatives of the new science.

In *Rambler* 24 the reader is reminded that science, like all studies, can be pursued to extremes. "Gelidus,"[42] a diligent and intelligent researcher, is dehumanized by his studies. Completely lacking human sympathy, Gelidus looks upon calamity and suffering and is reminded of his scientific concerns. When his family notifies him that a nearby town is being ravaged by fire which is spreading in all directions, Gelidus comments that their

42. According to Mrs. Thrale (*Anecdotes*, p. 179), the character of Gelidus was "meant to represent" John Colson (1680–1760), F.R.S., 1713, appointed Lucasian professor of mathematics at Cambridge in 1739, to whom Walmesley wrote on behalf of Johnson and Garrick. See *Life*, I, 101–3; James L. Clifford, *Young Sam Johnson* (New York, 1955), pp. 168, 170, 180, 182. Colson translated Musschenbroek's *Elementa Physicae*; Johnson cited the translation in the *Dictionary*. See W. K. Wimsatt, *Philosophic Words* (New Haven, 1948), pp. 30, 55, 156.

report is probably accurate, "for fire naturally acts in a circle" (*Rambler*, III, 133). Johnson's subject here is the common duties of life, not the reprehensible nature of scientific study. Primary responsibilities must receive the attention they demand; no pursuit should supersede our human obligations, but an affection for "philosophy" is only one of many things that may be mistakenly given the primary position in an individual's system of values. "Polyphilus," of *Rambler* 19, is extremely intelligent but never settles on a vocation or particular branch of study. He attempts to attain, in turn, scientific, legal, military, and linguistic eminence. The Rambler regrets the waste of energy and genius which results from a lack of staying power; the essay is a defense of diligence and early resolution. "Pertinax" (*Rambler* 95), whose studies include natural philosophy, is an extreme skeptic. He reaches the point of total confusion and mental distortion, for he has no method for proceeding to truth. He is finally saved by tolerating that which he cannot confute, refraining from the holding of false views for the sake of argument, and avoiding unanswerable questions.

Rambler 117 is a letter from "Hypertatus," a Swiftian spoof in which the correspondent argues the advantages of living in a garret on "scientific" principles. Johnson, of course, vigorously opposed the claims that external circumstances such as weather or climate alter or enhance a man's effectiveness as scholar or writer,[43] but this essay is delightful, not caustic. There are no damning thrusts, no outrage at the pretensions of this theorist, and, of course, no fears that the coming of chaos is at hand. It is an entertaining *jeu d'esprit*. The character of "Gelasimus" in *Rambler* 179 deals with a mathematician unacquainted with the proper modes of social intercourse who becomes a buffoon in order to please. It has nothing really to do with science, but shows that the compartmentalization of knowledge hinders con-

---

43. For an extended refutation of the principles of such men as Hypertatus, see *Idler* 11, but cf. Edward J. Van Liere, "Doctor Johnson and the Weather," *Philological Papers: University of West Virginia*, Series 52, No. 4–1 (Oct. 1951), pp. 40–48. Curtis Bradford has noted an analogue to *Rambler* 117 in an anonymous leaflet of 1751 cited in Wimsatt, *Philosophic Words*, p. 119, n. 40.

versation and joviality. In *Rambler* 199, "Hermeticus," a harmless projector who willingly becomes his own guinea pig in dangerous experiments, discourses on Rabbi Abraham Ben Hannase's account of the virtues of the magnet. Rabbi Abraham asserts that magnets will discover infidelity in women and Hermeticus outlines his plans for full-scale production of them in various shapes and strengths. The paper is light in tone and generous in spirit, and the interest shifts quickly from scientific to marital matters. In a sense, the paper involves an implied defense of scientific skepticism: Hermeticus is so credulous that he accepts the claims of Rabbi Abraham without question.[44]

Though science and amateur scientists are mentioned frequently in these papers, there is little that could be construed as serious satire or searching criticism of "philosophy." Johnson's specific comments are reserved for more favorable contexts, so that even a famous essay like the character of "Sober" in *Idler* 31 is at best peripheral in importance. Sober was, on Mrs. Thrale's authority,[45] a self-portrait, and there are, to be sure, some similarities between Johnson and his alleged counterpart—the pleasure in conversation, fear of solitude, and interest in chemistry—but Johnson, as any reader of the *Diaries, Prayers, and Annals* knows, is his own sternest critic. The portrait is not a genuine one. Sober avoids difficult projects; Johnson does not. Sober is a harmless dabbler, a would-be jack-of-all-trades who fails in his attempts; his creator is an inquisitive man, learned in the most diverse areas of human thought and experience. Literary psychoanalysts may find considerable interest in this self-depreciation, but the fact that great men often consider their works or intellectual attainments as poor things does not necessitate our sharing their opinions. To consider Johnson's interest in science as an offhand, trifling thing is surely erroneous; to associate his connection with scientific thought with the portrait of Sober is to patronize him, something that can never be ventured with impunity. Similarly, I see the astronomer of *Rasselas* as a psychological rather than a scientific portrait. The astronomer can-

44. For a parallel to *Rambler* 199 in Browne's *Pseudodoxia*, see ibid., p. 73.
45. *Anecdotes*, p. 178.

not prove his control of the weather and seasons "by any external evidence" and has fallen prey to an Idol of the Cave. The portrait enforces a Baconian reading of human fallibility and is not an assault on science. Madness and its general cause are at issue, not the individual's particular pursuit.[46]

The most important scientific "character" in the periodical essays is that of "Quisquilius" in *Rambler* 82, not because of the playful gibes directed at the Rambler's correspondent, but because it calls forth *Rambler* 83, a defense of curiosity and justification for the virtuoso's activities. In his letter Quisquilius proclaims himself "the most laborious and zealous virtuoso that the present age has had the honour of producing . . . ." As a child, he breaks his toys to "discover the method of their structure, and the causes of their motions . . . ." He delights in rummaging through old houses, wishes he had been alive during the time of the dissolution of the monasteries, and allows his tenants to pay their rents in butterflies, horseflies, grubs, or some similar specie. He has the longest blade of grass on record, a vial containing "dew brushed from a banana in the gardens of Ispahan," a snail that once crawled on the wall of China, and similar valuables. As might be expected, his purchases end in bankruptcy. He sends the Rambler a pebble picked up by Tavernier on the banks of the Ganges, in return for which he hopes that his sale catalogue will be recommended to the public. The extent of his absurdities renders him irresistible rather than blameworthy, and the portrait is even more interesting in that Quisquilius carefully follows several of the principles which the Rambler is at pains to recommend. Unlike "Polyphilus" he completes his projects. He is incredibly assiduous, has contempt for all narrow conceptions, and has studied past ages as well as present ones.

Rambler 83 sets forth most of the principles which Johnson

---

46. Interestingly enough, Sprat (*History of the Royal Society*, p. 342) argues "that the *Real Philosophy* will supply our thoughts with excellent *Medicines*, against their own *Extravagances* . . . ." Had the astronomer turned to experimental science he might have been saved. See also Kathleen M. Grange's important article, "Dr. Samuel Johnson's Account of a Schizophrenic Illness in *Rasselas* (1759)," *Medical History*, 6 (April 1962), 162–68, 291.

espouses in his justification of the virtuoso's curiosity and practices. He points out that many fields of inquiry are so remote from clearly useful knowledge and so ostensibly unimportant to human happiness or virtue that it is difficult to suppress our merriment or pity:

> Yet it is dangerous to discourage well-intended labours, or innocent curiosity; for he who is employed in searches, which by any deduction of consequences tend to the benefit of life, is surely laudable, in comparison of those who spend their time in counteracting happiness, and filling the world with wrong and danger, confusion and remorse (*Rambler*, IV, 71–72).

Harmless pleasure is always preferable to wickedness and debauchery—as Sprat had duly noted[47]—and what appears insignificant may prove useful: "It is impossible to determine the limits of enquiry, or to foresee what consequences a new discovery may produce. He who suffers not his faculties to lie torpid, has a chance, whatever be his employment, of doing good to his fellow-creatures."[48] The argument of physico-theology, that scientific study can be an inducement to piety, is invoked. The study of anatomy, for example, is important even in its most trifling form if it provides further reason for worshiping God: "there is nothing more worthy of admiration to a philosophical eye, than the structure of animals . . . and of all natural bodies it

47. *History of the Royal Society*, pp. 343–44. "And I dare challenge all the corrupt *Arts* of our *Senses*, or the devices of voluptuous wits, to provide fuller, more changeable, or nearer objects, for the contentment of mens *minds*. It were indeed to be wish'd, that severe virtu itself, attended only by its own *Authority*, were powerful enough to establish its dominion. But it cannot be so. The corruptions, and infirmities of *human Nature* stand in need of all manner of allurements, to draw us to good, and quiet manners. I will therefore propose for this end this cours of *Study*, which will not affright us with rigid praecepts, or sou'r looks, or peevish commands, but consists of sensible *pleasure*, and besides will be most lasting in its satisfaction, and innocent in its remembrance."

48. *Rambler*, IV, 72. Cf. ibid., no. 122, 286, where Johnson points out that the possible success of a given project is very difficult to estimate: "Nothing is more subject to mistake and disappointment than anticipated judgment concerning the easiness or difficulty of any undertaking . . . ."

must be generally confessed, that they exhibit evidences of infinite wisdom, bear their testimony to the supreme reason, and excite in the mind new raptures of gratitude, and new incentives to piety" (*Rambler*, IV, 72–73).

The performer Johnson, riding his three horses, demonstrates a new human capability, but the fact in itself is far less important than its inspirational value. The same point is made at length, in a "philosophic" context, in *Rambler* 83:

> To collect the productions of art, and examples of mechanical science or manual ability, is unquestionably useful, even when the things themselves are of small importance, because it is always advantageous to know how far the human powers have proceeded, and how much experience has found to be within the reach of diligence. Idleness and timidity often despair without being overcome, and forbear attempts for fear of being defeated; and we may promote the invigoration of faint endeavours, by shewing what has been already performed (*Rambler*, IV, 73).

The principle is an important one, not only because Johnson continually seeks to exhort and encourage, but because Bacon had attempted precisely the same thing; the "greatest obstacle to the progress of science and to the undertaking of new tasks and provinces therein, is found in this—that men despair and think things impossible."[49] Johnson also indicates the possibility that scientific serendipity may operate. Ingenuity may be exerted in a silly undertaking, but the principles and methods called forth by that ingenuity may be put to valuable purposes by another man.

The essay concludes with a discussion of the levels of scientific activity. It is unfortunate if the virtuoso works below his genius. He may, among his toys and trinkets, forget the necessity of meditation which leads to principles. Science demands the humble work of the persevering empiricist, but one must rise to the dignity of scientific generalization. For science to progress there must be labor. For it to be great, the labor must be allied with genius. But, and this is important, it may be thoroughly respectable and deserving of sympathy without being great.

Thus, after depicting with great wit the extremes of virtu-

49. Bacon, *Works*, IV (*The New Organon*), 90.

osity in *Rambler* 82, Johnson follows with an entire essay of justification and qualification. *Rambler* 177, which deals with a club of antiquaries, follows the same pattern.[50] After the letter of "Vivaculus" detailing the petty jealousy and ignorance besetting the society, Johnson adds a note of qualification: "It may, however, somewhat mollify his [Vivaculus'] anger to reflect, that perhaps, none of the assembly which he describes, was capable of any nobler employment, and that he who does his best, however little, is always to be distinguished from him who does nothing. Whatever busies the mind without corrupting it, has at least this use, that it rescues the day from idleness, and he that is never idle will not often be vitious" (*Rambler*, v, 172).

Johnson justifies the virtuoso's practices in several ways. Five have already been noted. The study of trifles is far better than idleness or wickedness, and may lead to important discoveries. Scientific study can enhance the virtuoso's sense of divine power and goodness. His work, if it proves useless in itself, can inspire others simply because he has worked so diligently, and if his methods or theories prove unsuccessful they may be useful to another, perhaps greater, researcher. Two other justifications are suggested. First, the collection of curiosities or observation of the natural world can prove to be a wholesome diversion, apart from its negative value of being preferable to evil activities or the idleness which usually precedes them:

Mankind must necessarily be diversified by various tastes, since life affords and requires such multiplicity of employments, and a nation of naturalists is neither to be hoped, or desired; but it is surely not improper to point out a fresh amusement to those who languish in health, and repine in plenty, for want of some source of diversion that may be less easily exhausted, and to inform the multitudes of both sexes, who are burthened with every new day, that there are many shows which they have not seen (*Rambler*, iii, no. 5, 29).

Johnson's pragmatic orientation toward science is extended to the individual. The trifler is prevented from debauchery; the serious student for whom science is merely an avocation can

50. Cf. Johnson's practice in *Rambler* 180, to be discussed in the following chapter.

use it therapeutically, to provide rest and relaxation after intense intellectual exertions: "The pride or the pleasure of making collections, if it be restrained by prudence and morality, produces a pleasing remission after more laborious studies . . . ."[51]

With this sense of proportion in hand, Johnson is impatient with the critics of the virtuoso and projector. If a man fails in a project or involves himself in folly, the happiness of others is not affected. Thus he says of Dodsley's *Preceptor*: "if it fails, nobody is hurt . . . ."[52] Johnson can see numerous ways in which seemingly useless or trifling activities may be justified. The scientific dabblers may not deserve eminence or fame, but they do not merit censure or derision.[53] After all, "to be able . . . to furnish pleasure that is harmless, pleasure pure and unalloyed, is as great a power as man can possess" (*Life*, III, 388).

Despite his sympathies for the harmless virtuoso, Johnson realizes that "a man would never undertake great things, could he be amused with small" (*Life*, III, 242). Thus he encourages the student of petty or trivial subjects to attempt greater things, and states that even rashness is preferable to cowardice: "It were to be wished that they who devote their lives to study would at once believe nothing too great for their attainment, and consider nothing as too little for their regard . . ." (*Rambler*, IV, no. 137, 362). Projectors must realize at the outset that their tasks will be difficult, but they should not hesitate to push ahead vigorously:

False hopes and false terrors are equally to be avoided. Every man, who proposes to grow eminent by learning, should carry in his mind,

51. *Idler* 56, p. 177. Cf. *Rambler*, IV, no. 85, 84–85: "when the mind is weary with its proper task, it may be relaxed by a slighter attention to some mechanical operation . . . ." (Johnson is here approving advice from Locke.)

52. *Prefaces & Dedications*, p. 188.

53. In *Rambler*, IV, no. 105, a dream vision in which the writer is guided by the goddess Curiosity, several absurd projectors appear, but this important qualification is made (p. 198): "Justice considered these projects as of no importance but to their authors, and therefore scarcely condescended to examine them; but Truth refused to admit them into the register." The projectors do not deserve fame but their work does not warrant damning criticism. Cf. his discussion of hack writers in *Rambler* 145. They deserve kindness but not reverence. Johnson is just but generous in spirit.

at once, the difficulty of excellence, and the force of industry; and remember that fame is not conferred but as the recompense of labour, and that labour, vigorously continued, has not often failed of its reward (*Rambler*, III, no. 25, 140).

In every human design there are dangers. In *Adventurer* 69 "projecting" is treated in terms of the vanity of human wishes theme. Hope is a sustaining force which must not be indulged to the point of delusion. Johnson distinguishes between the "hopes of folly" and the "hopes of reason." Hope may be a palliative but it must not become the ruling force in human motivation. It must invigorate but not intoxicate, and an appropriate test is prescribed, a means of distinguishing just plans and desires from fallacious or delusive ones:

To this test let every man bring his imaginations, before they have been too long predominant in his mind. Whatever is true will bear to be related, whatever is rational will endure to be explained: but when we delight to brood in secret over future happiness, and silently to employ our meditations upon schemes of which we are conscious that the bare mention would expose us to derision and contempt; we should then remember, that we are cheating ourselves by voluntary delusions; and giving up to the unreal mockeries of fancy, those hours in which solid advantages might be attained by sober thought and rational assiduity (*Adventurer*, p. 393).

Proposed projects must be reasonable, but again, as he notes in *Adventurer* 81, rashness is far more valuable than despondency:

he that dares to think well of himself, will not always prove to be mistaken, and the good effects of his confidence will then appear in great attempts and great performances; if he should not fully complete his design, he will at least advance it so far as to leave an easier task for him that succeeds him; and even though he should wholly fail, he will fail with honour.

But from the opposite error, from torpid despondency, can come no advantage; it is the frost of the soul which binds up all its powers, and congeals life in perpetual sterility. He that has no hopes of success will make no attempts, and where nothing is attempted, nothing can be done (*Adventurer*, p. 401).

In other words, the virtuoso should enlarge his vision and plans. What has proved useful to him should be seen in a wider and more sophisticated context, and, after arduous study, experimentation, and meditation, culminate in a work that is useful to all. "Nugaculus" of *Rambler* 103, who begins as a psychologist and moralist but degenerates into a snooping gossip, fails because his initial plan is incomplete. If a student is to avoid lapsing into folly or the study of trivialities, he must project something important at the outset: "The necessity of doing something, and the fear of undertaking much, sinks the historian to a genealogist, the philosopher to a journalist of the weather, and the mathematician to a constructer of dials" (*Rambler*, IV, 187).

It has been said that Johnson's "expressed view of science is on the whole melancholy . . . ."[54] Some of the remarks on virtuosi and projectors in the periodical essays are indeed concerned with the difficulties and follies attending certain quasi-scientific pursuits, but I would state the matter in different terms. Johnson's view of life is on the whole melancholy,[55] and since little can be expected from man, any attainment (except, of course, an "immoral" one) is worthy of admiration, and every reasonable proposal deserves encouragement. *Idler* 88, a discussion of the Royal Society, closes on the following note:

> From this mistaken notion of human greatness [that man too often lacks diligence rather than power] it proceeds, that many who pretend to have made great advances in wisdom so loudly declare that they despise themselves. If I had ever found any of the self-contemners much irritated or pained by the consciousness of their meanness, I should have given them consolation by observing, that a little more than nothing is as much as can be expected from a being who with respect to the multitudes about him is himself little more than nothing. Every man is obliged by the supreme master of the universe to improve all the opportunities of good which are afforded him, and to keep in continual activity such abilities as are bestowed upon him. But he has no reason to repine though his abilities are small and his opportunities few. He that has improved the virtue or advanced the happiness of one fellow-creature, he that has ascer-

54. Wimsatt, *Philosophic Words*, p. 56.
55. J. P. Hardy provides a series of references to Johnson's statements concerning the misery of human life. See *Rasselas*, p. 141. Cf. also 1825 *Works*, IX (*Sermon* XI), 392.

tained a single moral proposition, or added one useful experiment to natural knowledge, may be contented with his own performance, and, with respect to mortals like himself, may demand, like Augustus, to be dismissed at his departure with applause (*Idler*, pp. 274–75).

The lesson of *Idler* 88 is most poignantly articulated in Johnson's elegy on Robert Levet.[56] It is the lesson of Matthew 25:14–30, but Johnson's perception of human frailty allows God to apportion only one talent, not two or five, to his good and faithful servant. Human life and human abilities being what they are, there can be few Bacons, Boyles, and Newtons, but every man should, like the Levet of Johnson's elegy, contribute as much as he is able. Providence judges endeavors, not successes, and since the endeavors of the majority of men will fall far below those of the great "philosophers," Johnson gives them his sympathy and encouragement. It is the universality of their plight that captures his attention: not *their* lowliness and weakness, but the weakness of mankind.

56. Maurice Quinlan has also pointed out the parallel between *Idler* 88 and the Levet elegy. See *Samuel Johnson: A Layman's Religion* (Madison, 1964), pp. 25–26.

CHAPTER V

The Perpetual Moralist

In attempting to define the humanist temper by means of a listing of shared characteristics and principles, Paul Fussell observes that "The humanist is convinced that man's primary obligation is the strenuous determination of moral questions; he thus believes that inquiries into the technical operation of the external world ('science') constitute not only distinctly secondary but even irrelevant and perhaps dangerous activities. Johnson stresses the primacy of man's moral nature by insisting, in the *Life of Milton*, that 'We are perpetually moralists, but we are geometricians only by chance'."[1] Fussell is, of course, only one in a long line of commentators who have called attention to the famous dictum in the *Life of Milton*; it is perhaps Johnson's most familiar statement on science. As such, it deserves—and requires —considerable qualification. Actually, three passages are relevant here, the first two being from the 24th and 180th *Ramblers*, both of which bear on Johnson's statement in the *Life of Milton*:

1. *The Rhetorical World of Augustan Humanism* (Oxford, 1965), p. 7.

When a man employs himself upon remote and unnecessary subjects, and wastes his life upon questions, which cannot be resolved, and of which the solution would conduce very little to the advancement of happiness; when he lavishes his hours in calculating the weight of the terraqueous globe, or in adjusting successive systems of worlds beyond the reach of the telescope; he may be very properly recalled from his excursions by this precept ["know thyself"], and reminded that there is a nearer being with which it is his duty to be more acquainted; and from which, his attention has hitherto been withheld, by studies, to which he has no other motive, than vanity or curiosity.

The great praise of Socrates is, that he drew the wits of Greece, by his instruction and example, from the vain pursuit of natural philosophy to moral inquiries, and turned their thoughts from stars and tides, and matter and motion, upon the various modes of virtue, and relations of life (*Rambler*, III, no. 24, 131–32).

In *Rambler* 180 Raphael's advice to Adam in Book VIII of *Paradise Lost* is praised as it is in the *Life of Milton*. The essay closes with this remark:

If, instead of wandering after the meteors of philosophy which fill the world with splendour for a while, and then sink and are forgotten, the candidates of learning fixed their eyes upon the permanent lustre of moral and religious truth, they would find a more certain direction to happiness. A little plausibility of discourse, and acquaintance with unnecessary speculations, is dearly purchased when it excludes those instructions which fortify the heart with resolution, and exalt the spirit to independence (*Rambler*, V, no. 180, 186).

In the *Life of Milton*, the *locus classicus*, Johnson states that "Raphael's reproof of Adam's curiosity after the planetary motions, with the answer returned by Adam, may be confidently opposed to any rule of life which any poet has delivered" (*Lives of the Poets*, I, 177). He comments at length on Milton's stress of scientific study in his ideal educational scheme:

the truth is that the knowledge of external nature, and the sciences which that knowledge requires, or includes, are not the great or the frequent business of the human mind. Whether we provide for action or conversation, whether we wish to be useful or pleasing, the first requisite is the religious and moral knowledge [Johnson links the two] of right and wrong; the next is an acquaintance with the history of mankind, and with those examples which may be said to embody truth and prove by events the reasonableness of opinions. Prudence and Justice are virtues and excellences of all times and of all places; we are perpetually moralists, but we are geometricians only by chance. Our intercourse with intellectual nature is necessary; our speculations upon matter are voluntary and at leisure. Physiological learning is of such rare emergence that one man may know another half his life without being able to estimate his skill in hydrostaticks or astronomy, but his moral and prudential character immediately appears.

Those authors, therefore, are to be read at schools that supply most axioms of prudence, most principles of moral truth, and most materials for conversation; and these purposes are best served by poets, orators, and historians.

Let me not be censured for this digression as pedantick or paradoxical, for if I have Milton against me I have Socrates on my side. It was his labour to turn philosophy from the study of nature to speculations upon life, but the innovators whom I oppose are turning off attention from life to nature. They seem to think that we are placed here to watch the growth of plants, or the motions of the stars. Socrates was rather of opinion that what we had to learn was, how to do good and avoid evil (*Lives of the Poets*, ɪ, 99–100).

There can be no doubt as to the primacy of moral and religious duty in Johnson's conception of human life. In *Adventurer* 119 he writes of those who pursue trifling activities at the expense of important matters, men "who are kept from sleep by the want of a shell particularly variegated . . . who hover like vultures round the owner of a fossil, in hopes to plunder his cabinet at his death; and who would not much regret to see a street in flames, if a box of medals might be scattered in the tumult" (*Adventurer*, p. 464). The astronomer in *Rasselas* laments that he has passed his time "in the attainment of sciences which can, for the most part, be but remotely useful to mankind" (*Rasselas*, p. 113). Readers of the *Preceptor* are told that "other Acquisitions [than ethics or morality] are merely temporary Benefits, except as they contribute to illustrate the Knowledge, and confirm the Practice of Morality and Piety, which extend their Influence beyond the Grave, and increase our Happiness through endless Duration" (*Prefaces & Dedications*, p. 186). Religion is "the most important of all subjects..." (*Life*, III, 298). "In comparison of [securing happiness in another world], how little are all other things!" (*Life*, II, 358). Charity is more important than curiosity: "Of the divine author of our religion it is impossible to peruse the evangelical histories, without observing how little he favoured the vanity of inquisitiveness; how much more rarely he condescended to satisfy curiosity, than to relieve distress; and how much he desired that his followers should rather excel in goodness than in knowledge" (*Rambler*, IV, no. 81, 61).

The point at issue is not the preeminence of religion and morality, but the mutual exclusivity of what John Hardy calls "self-knowledge" and "star-knowledge."[2] In other words, does the pursuit of wisdom, the concentration on "various modes of virtue, and relations of life" constitute an activity so different from that of the natural philosopher, that man must choose the former to the exclusion of the latter? As will be seen, the elevation of man's moral and religious concerns involves no derogation of scientific knowledge or pursuits. On the contrary, Johnson attempts to demonstrate the ways in which science complements ultimate human concerns, instead of leading man astray from them. Why then, is he so firm and ostensibly final in the *Life of Milton*?

Robert Voitle judges the "pessimism" and "fatalism" of *The Vanity of Human Wishes* uncharacteristic of Johnson and argues that here, as in *The Vision of Theodore*, he "is exercising the moralist's license to overstate."[3] Though I do not concur in

2. See his "Johnson and Raphael's Counsel to Adam," in *Johnson, Boswell and their Circle* (Oxford, 1965), pp. 122–36, and *Rasselas*, pp. 132, 170, 175.

3. *Samuel Johnson the Moralist* (Cambridge, Mass., 1961), p. 45. Cf., with the praise of Socrates, the remark (*Life*, III, 265–66) that "were Socrates and Charles the Twelfth of Sweden both present in any company, and Socrates to say, 'Follow me, and hear a lecture in philosophy;' and Charles, laying his hand on his sword, to say, 'Follow me, and dethrone the Czar;' a man would be ashamed to follow Socrates." Overstatement can lead Johnson in the other direction as well.

this reading of *The Vanity of Human Wishes*, his point that Johnson, as moralist, may sometimes be willing purposely to overstate his case is well taken, and the reason for the rigor of the pronouncement in the *Life of Milton* is not far to seek. When William Seward wondered "that there should be people without religion," Johnson answered that the fact should cause little surprise "when you consider how large a proportion of almost every man's life is passed without thinking of it" (*Life*, IV, 215). Since man is all too prone to forget his moral and religious duties, Johnson feels compelled to remind him of them. In his tenth sermon, he discusses the absence of religious principle and lack of self-knowledge which characterize the actions of a great number of men:

> Many men may be observed, not agitated by very violent passions, nor overborn by any powerful habits, nor depraved by any great degrees of wickedness; men who are honest dealers, faithful friends, and inoffensive neighbours; who yet have no vital principle of religion; who live wholly without self-examination: and indulge any desire that happens to arise, with very little resistance or compunction; who hardly know what it is to combat a temptation, or to repent of a fault; but go on, neither self-approved, nor self-condemned; not endeavouring after any excellence, nor reforming any vitious practice, or irregular desire.[4]

I am also persuaded that the principles quoted from the *Life of Milton* and the *Ramblers* are exaggerated and overstated, for they conflict with Johnson's omnipresent encouragement of and admiration for curiosity, "in great and generous minds, the first passion and the last . . . ."[5] Quoting the statements apart from context has proven especially misleading here, for just as Johnson qualifies the harsh remarks on virtuosi in *Ramblers* 82 ("Quisquilius") and 177 ("Vivaculus" and the society of antiquaries), he adds a comment in *Rambler* 180 to his praise of Raphael's counsel which is far more characteristic and sets the matter in proper perspective:

4. 1825 *Works*, IX (*Sermon* x), 381.
5. *Rambler*, v, no. 150, 34. Hardy cites numerous passages in which Johnson praises curiosity in his "Johnson and Raphael's Counsel to Adam," pp. 122–24.

I am far from any intention to limit curiosity, or confine the labours of learning to arts of immediate and necessary use. It is only from the various essays of experimental industry, and the vague excursions of minds sent out upon discovery, that any advancement of knowledge can be expected; and though many must be disappointed in their labours, yet they are not to be charged with having spent their time in vain; their example contributed to inspire emulation, and their miscarriages taught others the way to success.

But the distant hope of being one day useful or eminent, ought not to mislead us too far from that study which is equally requisite to the great and mean, to the celebrated and obscure; the art of moderating the desires, of repressing the appetites; and of conciliating, or retaining the favour of mankind (*Rambler*, v, 183–84).

Johnson's considered judgment is that man's self-knowledge, his relation to God and his fellow men, are more important than scientific inquiry, but the question is one of relative position within a hierarchy of values, not of opposition between the "humanist" and scientific temperaments. Salvation, scientific learning, and the moral life in no way exclude one another. Johnson's statements are intended to remind, not to attack or harass; he is not presenting an either-or choice. His cautionary remarks to the scientist that his involved studies may increase the danger of forgetting more important concerns can coexist with his strong praise of scientific curiosity, methodology, and achievement without strain. Put simply, first things must come first. In comparison with the constant duty of man to fulfill his moral responsibilities and work toward his salvation, all human activities are small in importance. If science must take a second seat, poetry must do so also.[6]

As was briefly indicated earlier, the complementary relations

6. See, for example, *Life*, ii, 351, where Johnson says that poetry is "merely a luxury" or *Johnson on Shakespeare*, ed. Arthur Sherbo, 2 vols. (New Haven, 1968), vii, 102, where he points out that the subjects discussed by the literary scholar "are of very small importance . . . ." Cf. Walter Raleigh, *Six Essays on Johnson* (Oxford, 1910), p. 156: "Perhaps his chief difference from the critics of other schools is to be found in his comparatively low estimate of the importance of poetry; and this was due, not to any contempt, for he had been all his life a reader and lover of poetry, but to his deep sense of the greater issues of life and death."

among science, religion, and the moral life are far more interesting than any conflicting ones. We have already pointed out Johnson's concern with utilitarian applications of science; he follows Bacon in linking the relief of man's estate with the practice of Christian charity, and could easily subscribe to Bacon's judgment that knowledge should not be sought "either for pleasure of the mind, or for contention, or for superiority to others, or for profit, or fame, or power, or any of these inferior things; but for the benefit and use of life . . . ." The angels fell from lust of power, man from lust of knowledge, "but of charity there can be no excess, neither did angel or man ever come in danger by it."[7] Thus, in reviewing Stephen Hales's work on the distillation of sea water, ventilators in ships, and the curing of ill taste in milk, he says, "This is another of the labours of a life spent in the service of mankind."[8] The scientist is viewed as a Christian public benefactor; in the relief of human suffering he enjoys a unique role. This is one of the major reasons for Johnson's affection and respect for physicians, and the praise of Hales in this regard is surely justified. Besides his well-known work in plant physiology, his devising of the pneumatic trough for the collection of gases, and his discovery of variation of blood pressure in health and sickness, Hales also attempted to discover some method of ventilating buildings, particularly hospitals and prisons, without adding to the cost of the window tax. He eventually invented the windmill ventilator, which was installed in a number of prisons, with benefit to the inmates, whose death rate was reported to have been reduced because of Hales's device.[9] Clearly, science at its best offers some positive solutions to the problems that beset man. Thus, Johnson's remarks concerning the primacy of moral and religious concerns should be compared with his

7. *Works*, IV (*The Great Instauration*), 21.
8. *Literary Magazine*, 1, iii, 143.
9. A. Wolf, *A History of Science, Technology, and Philosophy in the Eighteenth Century*, 2nd ed., rev. D. McKie (London, 1952), pp. 346, 440, 473, 667. Hales, a foreign member of the French Academy, and holder of the Royal Society's Copley Medal, was a neighbor of Pope's and one of the two witnesses to his will. See Marjorie Nicolson and G. S. Rousseau, "*This Long Disease, My Life*" (Princeton, 1968), pp. 104, 107, 108.

statement, "It is our first duty to serve society, and, after we have done that, we may attend wholly to the salvation of our own souls" (*Life*, II, 10), but there is really no reason why one cannot do both simultaneously.

Johnson, as one might assume, sided with Samuel Clarke in his controversial correspondence with Leibniz (*Tour*, p. 256). Beyond pointing out the fact that Leibniz has stubbornly refused to read what Newton actually wrote concerning space as the *sensorium numinis*, Johnson does not provide a detailed rationale for his alignment with Clarke. However, it comes as no surprise. Clarke was one of Johnson's favorite, albeit heretical, theological writers,[10] and in the correspondence he is defending Newtonian science against the objections of a "systematizer." One claim in particular of Leibniz—whom Johnson terms "as paltry a fellow as I know" (*Tour*, p. 256)—must have perturbed Johnson as much as it did Clarke, namely the charge that Newtonian views had contributed to a decay of natural religion in England.[11] The religious cast of English science is one of its chief differentiating characteristics. While in France the English scientific methodology could sometimes be appropriated with little regard for its religious dimension, the English scientists, scientific apologists, and divines were generally agreed with the psalmist that the heavens declare the glory of God, and he who studies the heavens, the earth, and the creatures that inhabit it, should be led to greater devotion. Browne was uttering a commonplace when, in *Religio Medici*, he asserted that "The wisedome of God receives small honour from those vulgar heads, that rudely stare about, and with a grosse rusticity admire his workes; those highly magnifie him whose judicious enquiry into

10. Johnson is continually reading Clarke. See *Samuel Johnson: Diaries, Prayers, and Annals*, ed. E. L. McAdam, Jr., with Donald and Mary Hyde (New Haven, 1958), pp. 105, 122, 129, 132, 155, 159, 305, 415. On Johnson and Clarke, see Maurice Quinlan, *Samuel Johnson: A Layman's Religion* (Madison, 1964), pp. 27–45.

11. H. G. Alexander, ed., *The Leibniz-Clarke Correspondence: Together with Extracts from Newton's "Principia" and "Opticks"* (Manchester, 1956), p. 11, and for Alexander's comment, p. xv.

his acts, and deliberate research of his creatures, returne the duty of a devout and learned admiration."[12] Browne's collecting of "Divinity" from the book of Nature as well as scripture is extremely representative. Bacon had indicated that the creation would open "our belief, in drawing us into a due meditation of the omnipotency of God, which is chiefly signed and engraven upon his works."[13] Sprat sees the admiration of the Creator through a study of His works as "the utmost perfection of *humane Nature*,"[14] and the gallery of physico-theological writers continually proclaims the magnificence of the natural world and the manner in which its complexity, plenitude, and above all, design, testify to the wisdom, power, and grand creativity of God. The linking of scientific discovery with theological principle that the physico-theological writers effected, and especially their stress of the existence and attributes of God which may be demonstrated from a consideration of the design of His creation, constitute one of the major, if not the major, characteristics of pre-Darwinian English science.[15]

Johnson's sympathy with the goals of the physico-theologians counterbalances his comment in the *Life of Milton*. The knowledge of external nature could well be the frequent business of the human mind: to a very important extent, we *are* placed here to watch the growth of plants, or the motions of the stars. He

12. Browne, *Works*, I, 22.

13. Bacon, *Works*, III (*The Proficience and Advancement of Learning*), 301.

14. Thomas Sprat, *History of the Royal Society*, ed. Jackson I. Cope and Harold Whitmore Jones (St. Louis, 1966), p. 111.

15. The best introduction to the historical role of physico-theology which I know is Charles E. Raven, *Natural Religion and Christian Theology*, vol. I, *Science and Religion* (Cambridge, 1953). The tradition is also discussed in Basil Willey, *The Eighteenth Century Background: Studies on the Idea of Nature in the Thought of the Period* (London, 1940); W. P. Jones, *The Rhetoric of Science* (London, 1966); and Alan Dugald McKillop, *The Background of Thomson's "Seasons"* (Minneapolis, 1942), but not in the detail which Raven provides. See also Raven, *John Ray, Naturalist: His Life and Works*, 2nd ed. (Cambridge, 1950). Though its focus is slightly different, one can also recommend Michael Macklern, *The Anatomy of the World: Relations between Natural and Moral Law from Donne to Pope* (Minneapolis, 1958).

praises the oration in which Boerhaave "asserts the Power and Wisdom of the Creator, from the wonderful Fabric of the human Body . . ."[16]—a common notion in the physico-theological works of such writers as Richard Bentley, John Ray, and Sir Richard Blackmore—and in the *Rambler* says that "a man that has formed this habit of turning every new object to his entertainment, finds in the productions of nature an inexhaustible stock of materials upon which he can employ himself. . . ," Such a man "has always a certain prospect of discovering new reasons for adoring the sovereign author of the universe . . ." (*Rambler*, III, no. 5, 28–29). Eighteenth-century physico-theological verse is often undistinguished, but Johnson points out one merit which it enjoys; it is one of the few forms in which piety can be properly motivated.

Let no pious ear be offended if I advance, in opposition to many authorities, that poetical devotion cannot often please. The doctrines of religion may indeed be defended in a didactick poem, and he who has the happy power of arguing in verse will not lose it because his subject is sacred. A poet may describe the beauty and the grandeur of Nature, the flowers of the Spring, and the harvests of Autumn, the vicissitudes of the Tide, and the revolutions of the Sky, and praise the Maker for his works in lines which no reader shall lay aside. The subject of the disputation is not piety, but the motives to piety; that of the description is not God, but the works of God.[17]

In discussing Johnson's frequent citation of physico-theological writers in the *Dictionary*, W. K. Wimsatt comments that "As John Ray had seen the wonders of God everywhere in the Creation, Johnson's readers found the same wonders everywhere in the variegated realm of discourse of which his Dictionary was the alphabetized reflection."[18] In the *Life of Milton* he suggests that students should read poets, orators, and historians, but in his Preface to Dodsley's *Preceptor*, which Boswell called

---

16. "Boerhaave," *Med. Dict.*, I, sig. 9U2ᵛ.

17. *Lives of the Poets*, I (*Life of Waller*), 291.

18. Wimsatt, "Johnson's Dictionary," in *New Light on Dr. Johnson*, ed. Frederick W. Hilles (New Haven, 1959), p. 83. See also A. D. Atkinson, "Dr. Johnson and Some Physico-Theological Themes," *Notes and Queries*, 197 (5 Jan., 12 April, 7 June, 1952), 16–18, 162–65, 249–53.

"one of the most valuable books for the improvement of young minds that has appeared in any language" (*Life*, I, 192), Johnson advises his youthful readers to peruse the literature of physicotheology:

To excite a Curiosity after the Works of God, is the chief Design of the small Specimen of *Natural History* inserted in this Collection; which, however, may be sufficient to put the Mind in Motion, and in some measure to direct its Steps; but its Effects may easily be improved by a Philosophic Master, who will every Day find a thousand Opportunities of turning the Attention of his Scholars to the Contemplation of the Objects that surround them, of laying open the wonderful Art with which every Part of the Universe is formed, and the Providence which governs the Vegetable and Animal Creation. He may lay before them the *Religious Philosopher* [of Nieuwentijdt], *Ray, Derham's Physico-Theology*, together with the *Spectacle de la Nature* [of Pluche] ....[19]

Johnson's most famous praise of such writings is his defense of Blackmore's *Creation*. He argues that the wits exposed Blackmore "to worse treatment than he deserved...."[20] Of Blackmore's poem he writes:

Not long after (1712) he published *Creation, A philosophical Poem*, which has been by my recommendation inserted in the late collection.[21] Whoever judges of this by any other of Blackmore's performances will do it injury. The praise given it by Addison (*Spec.* 339) is too well known to be transcribed; but some notice is due to

19. *Prefaces & Dedications*, pp. 185–86. Hazen, pp. 172–73, suggests a "more than casual" connection with the editing of *The Preceptor*, since Johnson, besides contributing the Preface and *The Vision of Theodore*, extensively revised the Preface for the second edition (June 1754). He also revised his lists of recommended reading for this edition, and I think we can say that he took his task quite seriously; inclusion in the lists indicates strong admiration on his part. Cf. *Tour*, p. 313, where, on Inchkenneth, Johnson chooses Derham's *Physico-Theology* for Sunday reading.

20. *Lives of the Poets*, II (*Life of Blackmore*), 252.

21. Johnson recommended the inclusion of Blackmore, Watts, Pomfret, and Yalden. See *Life*, III, 370–71, and *Lives of the Poets*, III (*Life of Watts*), 302. Some of Thomson's works were also suggested for inclusion. See *Letters*, II, no. 515 (3 May 1777 to Boswell), 170–71.

the testimony of Dennis, who calls it a 'philosophical Poem, which has equalled that of Lucretius in the beauty of its versification and infinitely surpassed it in the solidity and strength of its reasoning.'[22]

This is indeed high (and to those who have read *Creation*, questionable) praise. Johnson's judgment of the *Essay on Man* is far less sympathetic.

However, despite the poetic effusions of such writers as Blackmore and the nearly endless series of treatises demonstrating the warm relations existing between science and religion, the tradition of physico-theology was not without serious problems. In the first place, there was a tendency to see more relations, more design, more harmony, more law than experience could justify. Though the Christian virtuoso's approach to nature was empirical, his interpretation of his observations and experiments was fitted into a framework which stood on an a priori foundation. He was prepared to find natural laws before he actually did. Convinced that God had created a universe of peace and harmony, he perceived nature's beauties but often missed or ignored its ugliness or violence. As science was intended to underline revelation, prior belief could well influence science.[23] Thus Voltaire complains, "One writer wants to persuade me by means of physics to believe in the Trinity; he tells me the three persons of the deity are like the three dimensions of space. Another claims he will give me palpable proof of transubstantiation and shows me through the laws of motion how an accidental property can exist without its subject."[24] This tendency toward highly questionable rationalization on the part of professed, diffident empiricists is evident in Sprat, who calls miracles "*Divine Experiments*," Christ's feeding the hungry, curing the lame and blind "*Philosophical Works*,"[25] and considers Adam, "when he obey'd [God] in mustring, and

22. *Lives of the Poets*, II (*Life of Blackmore*), 242–43.

23. See Richard S. Westfall, *Science and Religion in Seventeenth-Century England* (New Haven, 1958), p. 69.

24. Cited by Ernst Cassirer, *The Philosophy of the Enlightenment*, trans. Fritz C. A. Koelln and James P. Pettegrove (Boston, 1955), p. 48.

25. *History of the Royal Society*, p. 352.

naming, and looking into the *Nature* of all the *Creatures*," the first natural philosopher and first student of natural religion.[26]

The ramifications of natural theology were not so salutary as the "philosophers" might have hoped. Though it was intended to enhance Christianity by buttressing its rational foundation, in practice it sometimes tended to displace the entire edifice.[27] Deism, in attempting to subscribe to a limited number of universal articles arrived at by reason with no recourse to revelation, is, by definition, non-Christian,[28] but the scientists, though preaching Christianity, concentrated on natural theology to the extent that their very Christianity became reduced to a handful of principles, consisting of veneration for God, obedience to the natural law, and belief in an afterlife. To an important degree, the scientists did more to challenge traditional Christianity than the legion of "atheists" against whom their treatises were consistently written. In their framework, the spiritual quest and growth of the individual soul becomes subordinate to the worship of an external God possessing titanic power.[29] Johnson, it need hardly be said, does not succumb to this tendency. He never loses sight of the preeminent position of divine revelation. In light of the comments of some students of Johnson, it has been inferred that "if Johnson let himself go, he might well turn into a deist."[30] Such a judgment, in my opinion, is incredible, but it is true that the delineation of personal belief is particularly difficult in the eighteenth century. The varying modes and forms of belief do shade into one another; there is considerable truth in the proverbial judgment that the sense of reduction and consolidation which one encounters in natural theology, and which

26. Ibid., pp. 349–50. See also R. F. Jones, *Ancients and Moderns*, 2nd ed. (St. Louis, 1961), pp. 231–32.

27. Westfall, *Science and Religion in Seventeenth-Century England*, p. 106.

28. For a clear and comprehensive discussion of Deism, see Phillip Harth, *Contexts of Dryden's Thought* (Chicago, 1968), pp. 58–94, *et passim*.

29. Westfall, *Science and Religion in Seventeenth-Century England*, pp. 141–42.

30. Voitle, *Samuel Johnson the Moralist*, p. 175. This is not Voitle's judgment, but a conclusion drawn from the approaches of others.

appears by definition in Deism, did combine with a search for agreement, compromise, harmony, and simplification in the aftermath of the sectarian turmoil of the preceding century to produce a religious atmosphere which may, at times, deservedly be termed bland or lifeless.

A final difficulty with the physico-theological tradition, one of which we can be sure Johnson was painfully aware, is the basic problem involved in the use of the argument from design and its reliance on teleology—the problem of the existence of evil.[31] As he announces at the outset of his review of Soame Jenyns' *Free Enquiry*, Johnson sees the problem as insoluble: "This is a treatise, consisting of six letters, upon a very difficult and important question, which, I am afraid, this author's endeavours will not free from the perplexity which has entangled the speculatists of all ages, and which must always continue while *we see* but *in part* . . . . he decides, too easily, upon questions out of the reach of human determination, with too little consideration of mortal weakness, and with too much vivacity for the necessary caution." In the course of this review Johnson also (like Voltaire)[32] takes issue with one of the most potent concepts which the physico-theological writers employed, that of the great chain of being.[33]

Nevertheless, as Chester F. Chapin points out, Johnson no doubt approved of the physico-theologians' arguments for theism.[34] His very citation of their works in the *Dictionary* and his recommendation of them in the Preface to the *Preceptor* indicates his sympathy; he takes his duty and responsibility in these roles very seriously. I consider it equally certain, however, that Johnson was aware of the limitations of physico-theology. Because his own religious life was so intense and so marked by

31. For an excellent discussion of the design argument in its eighteenth-century context, see Robert H. Hurlbutt, *Hume, Newton, and the Design Argument* (Lincoln, 1965).

32. Arthur O. Lovejoy, *The Great Chain of Being: A Study of the History of an Idea* (Cambridge, Mass., 1936), p. 184.

33. On the use of the great chain metaphor and other related ideas by literary figures influenced by physico-theology, see Jones, *The Rhetoric of Science*, pp. 28–30.

34. *The Religious Thought of Samuel Johnson* (Ann Arbor, 1968), p. 78.

struggle, uneasiness, and doubt,[35] he welcomed all additional "proof" for the existence of God or manifestations of the divine attributes, but he could never embrace natural theology to the extent that revelation would diminish in importance, for it is scripture which offers suffering man a partial answer to the problem of evil:

> You have demonstration for a First Cause: you see he must be good as well as powerful, because there is nothing to make him otherwise, and goodness of itself is preferable. Yet you have against this, what is very certain, the unhappiness of human life. This, however, gives us reason to hope for a future state of compensation, that there may be a perfect system. But of that we were not sure, till we had a positive revelation (*Life*, III, 316–17).

Natural theology was intended to complement revealed Christianity. If it became for some an alternative, it surely did not in Johnson's case, and in all probability he considered the dangers to be remote and not important enough to warrant his not recommending such writings. On the other hand, though he praised such works he could not accept their arguments as ironclad. I concur in the judgment of Hagstrum and Quinlan that Sermon XXI is not Johnson's; the wholesale acceptance in that sermon of arguments concerning the "adjustment, proportion, and accommodation of all matters in the wide creation" strikes me as

35. Chapin argues that Johnson "was neither basically a religious sceptic on the one hand nor a perfectly confident believer on the other. He was rather within the mainstream of that long and honorable Christian tradition in which a measure of doubt or 'uneasiness' is an element in the very composition of that ultimate commitment which is termed 'faith.'" See "Johnson and the 'Proofs' of Revelation," *Philological Quarterly*, 40 (April 1961), 301, and *The Religious Thought of Samuel Johnson*, pp. 177–78, n. 1. Quinlan's judgment is similar. See *Samuel Johnson: A Layman's Religion*, p. 175: "Johnson's faith was obviously not built on any simple plan. It was a blend of many things. Although based on hope and the promises of Scripture, although supported by traditional interpretations of the pre-Reformation church and the Church of England, it was tempered by what seemed to him reasonable in a logical sense or, in a few instances, credible according to empirical considerations. He probably did not accept any religious belief without grave and prolonged consideration, and thinking sometimes raised difficulties."

un-Johnsonian.[36] Such arguments, however, could buttress faith, and Johnson—like so many of his contemporaries—surely believed that new scientific discoveries had dramatized the divine attributes, and should be enlisted in the service of religion. Science does, in other words, enjoy an important relationship with other human concerns. One need not choose a moral or religious realm of action to the exclusion of scientific study. Their relationship is complementary. Unfortunately, however, science does not hold the answers to the most imposing of questions and the inferences which can be made on the basis of scientific discovery are not absolutely certain. One would be surprised indeed if Johnson was either completely convinced by or completely opposed to the arguments of the physico-theologians. Darwin, who is often popularly credited with finally undermining the argument from design, simultaneously perceived sufficient order and misery that he was forced to admit uncertainty and frustration.[37] Johnson's straightforward judgment that the

36. See ibid., p. 98. Quinlan notes the "approving allusions to the Chain of Being, a theory with which Johnson expressed strong disagreement elsewhere," and raises other objections. Though Johnson can certainly describe the complex series of physical relations which, if slightly altered, would bring catastrophe (marked increases or decreases in the earth's quantity of water, in the earth's proximity to the sun, etc.) and the design manifest in the parts of the human body, I would expect him to comment in some way on floods, earthquakes, diseases, and similar matters which one could expect to be on the minds of the attending congregation. Johnson, of course, had opposed the confident optimism of Pope and Soame Jenyns, accusing them of inexperience in the matter of human misery.

37. In his famous letter to Asa Gray, Darwin writes: "There seems to me too much misery in the world. I cannot persuade myself that a beneficent and omnipotent God would have designedly created the Ichneumonidae with the express intention of their feeding within the living bodies of caterpillars, or that a cat should play with mice. Not believing this, I see no necessity in the belief that the eye was expressly designed. On the other hand, I cannot anyhow be contented to view this wonderful universe, and especially the nature of man, and to conclude that everything is the result of brute force. I am inclined to look at everything as resulting from designed laws, with the details, whether good or bad, left to the working out of what we may call chance. Not that this notion at all satisfies me. I feel most deeply that the whole subject is too profound for the human intellect. A dog might as well speculate on the mind of Newton. Let each man hope

problem of evil is insoluble but the design and grandeur of the Creation inspiring, probably represents a consensus view. We are by now accustomed to seeing sophistication and complexity that, to adopt a Johnsonian phrase, would dare skepticism into silence; at the same time we continue to observe natural phenomena which, to human eyes at least, remain brutal and perplexing.

Like some of the early commentators on the Royal Society, Johnson has some fear that cavalier attitudes toward religion may follow on the heels of scientific discovery. He does, as will be seen directly, discuss men who have joined learning with piety, and he sees the union of scientific knowledge and religious fervor as a kind of ideal. He still, nevertheless, reminds the scientists that a reverent frame of mind should never be forsaken. Iconoclasm, smugness, and fashionable religious skepticism are reprehensible. He completely revised Anna Williams' "On the Death of Stephen Grey, F.R.S.," a cosmic voyage poem in the tradition of such works as Thomson's "Poem Sacred to the Memory of Sir Isaac Newton." In his revision Johnson indicates the limitations with which science, like all earthly pursuits, is beset, and points out that the scientist, to be blessed, must retain his Christian reverence:

> Now, hoary sage, pursue thy happy flight,
> With swifter motion haste to purer light,
> Where Bacon waits with Newton and with Boyle
> To hail thy genius, and applaud thy toil;
> Where intuition breaks through time and space,
> And mocks experiment's successive race;
> Sees tardy science toil at nature's laws,
> And wonders how th'effect obscures the cause.
>   Yet not to deep research or happy guess
> Is ow'd the life of hope, the death of peace.
> Unblest the man whom philosophick rage
> Shall tempt to lose the Christian in the sage;
> Not art but goodness pour'd the sacred ray
> That cheer'd the parting hour of humble Grey.
>   (*Poems*, 355, ll. 9–22)

and believe what he can." Cited in John C. Greene, *Darwin and the Modern World View* (New York, 1963), p. 44.

In his last illness his deep affection for Richard Brocklesby prompted him to seek assurance that the physician "should not entertain any loose speculative notions, but be confirmed in the truths of Christianity. . . ." An account was drawn up of the conversation which passed on the subject and Brocklesby signed it, Johnson urging him to "keep it in his own custody as long as he lived" (*Life*, IV, 414). Here there is no suggestion that science should not be studied vigorously, but the possible danger of freethinking is always to be avoided. Scientific study should enhance, not replace, the worship of God.

The passages in which Johnson finds himself in conflict with the new philosophy are very few, but he does take science to task when it conflicts with Genesis. He praises the *Philosophical Transactions*, "valuable collections, which have done so much honour to the *English* nation," but warns its contributors to avoid irreverence, complaining that in its pages "the sacrosanctity of religion . . . seems treated with too little reverence when it is represented as hypothetical and controvertible, that all mankind proceeded from one original."[38] For him the principle of evolution would have to be carefully qualified, but his apparent fundamentalism does not call into question his basic sympathy for empirical science.

In a very real sense the theoretical basis for scientific and religious harmony is of secondary or tertiary importance. Johnson is concerned with articulating it and does point out that the relief of human suffering is a Christian act facilitated by scientific findings, that the Creation illuminates God's glory and thus invites human investigation. Nevertheless, the strain of utility, as important in Johnson as it was in Bacon, tends to undercut the ratiocination and lead Johnson from the theoretically valid to the actual. The question at issue is not that of the possible relations between science and religion but the real ones: Are there pious scientists? Has their science impeded their piety? The

38. *Literary Magazine*, 1, iv (Review of *Philosophical Transactions*, XLIX, pt. 1), 193. Hardy, "Johnson and Raphael's Counsel to Adam," p. 127, n. 2, points out that Johnson is alluding to Henry Baker, "*A Supplement to the Account of a distempered Skin, published in the 424th Number of the* Philosophical Transactions," *Philosophical Transactions* for 1755, p. 23.

logical extension of Johnson's subscribing to the methodology of the new philosophers is to subject the physico-theologians' claims to the empirical test and ascertain whether or not there are men whose complex scientific studies coexist with and complement deep and sincere piety. He is interested in the Robert Levets and Stephen Grays, not in the abstractions that inhabit tracts and treatises. Thus, if we are to obtain his considered judgment we must seek it in the medical biographies, for biography—being both hortatory and efficacious—comes home to men's business and bosoms in a unique way: "Biography is, of the various kinds of narrative writing, that which is . . . most easily applied to the purposes of life" (*Idler* 84, p. 261). ". . . I esteem biography, as giving us what comes near to ourselves, what we can turn to use" (*Tour*, p. 55).

One of the most interesting aspects of the medical biographies is the extent to which theological or sectarian differences are minimized. Johnson is concerned with love of God and love of man; the heterodoxy of Browne or the puritanism of Sydenham are unimportant. The *Life of Browne* closes with a defense of the Norwich physician's essential Christianity. Johnson draws on all his powers in this defense and it is extremely impressive. One paragraph will indicate the direction of his argument:

> Whether Browne has been numbered among the contemners of religion, by the fury of its friends, or the artifice of its enemies, it is no difficult task to replace him among the most zealous professors of christianity. He may, perhaps, in the ardour of his imagination, have hazarded an expression, which a mind intent upon faults may interpret into heresy, if considered apart from the rest of his discourse; but a phrase is not to be opposed to volumes; there is scarcely a writer to be found, whose profession was not divinity, that has so frequently testified his belief of the sacred writings, has appealed to them with such unlimited submission, or mentioned them with such unvaried reverence (1825 *Works*, vi, 502).

It is noteworthy that Johnson devotes so much attention to this defense. In Johnson's judgment, Browne's faith is more important (for Browne—not necessarily for intellectual history) than his natural philosophy. But, equally important, the emphasis on religious questions does not lead Johnson to derogate lesser, scientific ones or even minute, antiquarian ones. Johnson seeks a realistic moral perspective, but no mutually exclusive choice.

In the *Life of Sydenham* Johnson also raises the issue of the pious scientist. In its closing paragraph, he suggests that Sydenham's benevolence, charity, and reverence make him a kind of paragon who should be adopted as a model by those who would follow him in the study of medicine:

> What was his character, as a physician, appears from the treatises which he has left, which it is not necessary to epitomise or transcribe; and from them it may likewise be collected, that his skill in physick was not his highest excellence; that his whole character was amiable; that his chief view was the benefit of mankind, and the chief motive of his actions, the will of God, whom he mentions with reverence, well becoming the most enlightened and most penetrating mind. He was benevolent, candid, and communicative, sincere, and religious; qualities, which it were happy, if they could copy from him, who emulate his knowledge, and imitate his methods (1825 *Works*, vi, 412).

Johnson's view is both pointed and clear: a man's accomplishments should benefit others, but greatness in the man is something higher, something beyond his works. Raleigh writes that "Johnson was great by his reserves; the best of him was withheld from literature; his books were mere outworks."[39] Raleigh's judgment is in a sense misleading; one is not prepared to accept it wholeheartedly for it is not grounded in a thorough appreciation of the importance of Johnson's writings. Nevertheless, Johnson would agree with the underlying principle.

The perception of a significant relation between divine revelation and human scientific discovery is all the more admirable in that it may be extremely difficult. Johnson applauds Newton's wrestling with these matters. Reviewing the Newton-Bentley Correspondence, which concerned Bentley's use of Newtonian discoveries in his Boyle lectures, he comments: "The principal question of these letters gives occasion to observe how even the

---

39. *Six Essays on Johnson*, p. 34.

mind of *Newton* gains ground gradually upon darkness."[40] Thus, like Sydenham and Browne, Newton, who "came to be a very firm believer" (*Life*, I, 455), earns his admiration, but his grand example of the pious scientist is Hermann Boerhaave. Johnson invariably mentions him with praise,[41] and his biography of the Dutch physician, chemist and botanist points up the dual nature of Boerhaave's eminence: the scientific achievement coupled with a deep religious awareness.

As was indicated earlier, Johnson's "scientific education" cannot be traced in satisfactory detail, but we can say that the outline and direction of his attitudes were formed early, for the *Life of Boerhaave*—written in his thirtieth year—includes nearly all of the principles which have been discussed in these pages. It provides a brief but full index to all of his major scientific attitudes. Though Johnson was acquainted with the works and personalities of greater scientists, it is perhaps not too rash to say that throughout his life Boerhaave remained his scientific model. The attitudes toward science which he expressed in 1739 never changed, and he never hesitated, throughout his career, to single out Boerhaave for special praise.

In this, the first of his medical biographies, Johnson praises the empirical, skeptical methodology; he argues the necessity of studying one's predecessors but also that one must distrust authority and judge for oneself. Though wary of premature or unfounded hypotheses, the scientist must rise to principle and synthesize the results of his experiments. Projectors are encouraged, detractors disparaged, physico-theology praised, harmless pleasure justified. Stylistic elegance in scientific writing is demanded. In the course of the biography Johnson reveals his sympathy with the scientific method by carefully evaluating the evidence before him and refusing to admit anything about which he is unsure. Besides being an inclusive index to his scien-

40. *Literary Magazine*, 1, ii, 89. Contrast Johnson's somber generalization with the enthusiastic uses to which Pope had put the light-darkness imagery in his "Epitaph. Intended for Sir Isaac Newton, In Westminster-Abbey."

41. See, for example, *Life*, II, 372; *Lives of the Poets*, II (*Life of Savage*), 334; *Adventurer* 85, p. 414; *Rambler*, IV, no. 114, 242.

tific attitudes, the biography points up his own "skeptical" practices.[42]

It has been suggested, justly I believe, that Boerhaave was a kind of hero to the young Johnson; that, to a large extent, he saw himself in this great man.[43] There are several notable parallels. Boerhaave was subjected to numerous forms of physical suffering, but rose above them, refusing to capitulate. Like Johnson he began in poverty, "but with a Resolution equal to his Abilities, and a Spirit not to be depress'd or shaken, he determin'd to break thro' the Obstacles of Poverty, and supply by Diligence the want of Fortune."[44] Like his biographer, Boerhaave was catholic in his interests, and exhibited a modest firmness in the company of great men. The parallel in physical characteristics is striking:

He was of a robust and athletic Constitution of Body, so hardened by early Severities, and wholsome Fatigue, that he was insensible of any Sharpness of Air, or Inclemency of Weather. He was tall, and remarkable for extraordinary Strength. There was in his Air and Motion something rough and artless, but so majestic and great

42. References to the *Med. Dict.* are in a sense pointless since the large folio leaves reduce the biography to a handful of pages and the reader must either cast about endlessly or pencil in paragraph numbers to follow my references. In the 1825 edition, VI (which follows the *Gentleman's Magazine* version), references are as follows: methodology: 275, 281; predecessors: 275–76, 289; authority: 272, 276; hypotheses: 281; synthesis: 280, 289; projectors: 275; detractors: 277, 285; physico-theology: 284; harmless pleasure: 271; stylistic elegance: 283, 289; evaluating evidence: 270, 285.

43. W. K. Wimsatt, *The Prose Style of Samuel Johnson* (New Haven, 1941), p. 123; James L. Clifford, *Young Sam Johnson* (New York, 1955), pp. 254–55.

44. "Boerhaave," *Med. Dict.*, I, sig. 9U1v. Dr. Morin's "unassisted merit" advanced slowly, but he too eventually triumphed. See 1825 *Works*, VI (*Life of Morin*), 393. Johnson's own feelings in his early years in the capital are thrown into bold relief and Morin can be added to the list of early heroes whose successes bolstered the confidence of the young Johnson. Cf. *Johnson on Shakespeare*, VII, 88–89: "Shakespeare . . . came to London a needy adventurer. . . . The genius of Shakespeare was not to be depressed by the weight of poverty, nor limited by the narrow conversation to which men in want are inevitably condemned; the incumbrances of his fortune were shaken from his mind, 'as dewdrops from a lion's mane.' "

at the same time, that no Man ever looked upon him without Veneration, and a kind of tacit Submission to the Superiority of his Genius.[45]

The strongest and most important parallel, however, lies in
the piety which, with Boerhaave, goes hand in hand with the
study of science and the practice of medicine. Two ways of life
competed for Boerhaave's attention: that of the physician and
that of the clergyman. His father had intended that he would
become, like himself, a minister, and Boerhaave resolved to do so;
but he chose medicine because, Johnson tells us, he was publicly
maligned as a disciple of Spinoza, a charge tantamount to atheism.
In 1690 he had "discussed the important and arduous Question
of the distinct Natures of the Soul and Body, with such Accuracy, Perspicuity, and Subtilty, that he entirely confuted all the
Sophistry of *Epicurus*, *Hobbes*, and *Spinosa*, and equally raised
the Character of his Piety and Erudition,"[46] but three years later
an event occurred whose repercussions formed an obstacle to
a religious career.

Sitting in a "common Boat" Boerhaave listened to a would-be
zealot declaim against Spinoza. The speaker proposed no arguments or refutations but restricted himself to invective and
personal abuse. When asked by Boerhaave if he had read Spinoza,
the man remained silent and "in a few Days, it was the common Conversation at *Leyden*, that *Boerhaave* had revolted to
Spinosa."[47] His friends pleaded on his behalf, pointing to "his
learned and unanswerable Confutation of all atheistical Opinions, and particularly of the System of *Spinosa*, in his Discourse
of the Distinction between Soul and Body," but the damage
had already been done:

> *Boerhaave*, finding this formidable Opposition raised against his
> Pretensions to Ecclesiastical Honours or Preferments, and even
> against his Design of assuming the Character of a Divine, thought
> it neither necessary nor prudent to struggle with the Torrent of
> popular Prejudice, as he was equally qualified for a Profession, not
> indeed of equal Dignity or Importance, but which must undoubtedly

45. *Med. Dict.*, I, sig. 9X1r.
46. Ibid., sig. 9U1v.
47. Ibid, sig 9U2r.

claim the second Place among those which are of the greatest Benefit to Mankind.[48]

Though not a clergyman, his extraordinary piety continued throughout his life. Johnson stresses this at length, and Boerhaave's learning and piety counterpoint one another in the organization of the biography.[49] The two elements constitute complementary motifs to which the incidents of the life are related. The very organization suggests that, in a scientific model, knowledge of the external world and worship of its Creator are inseparable. Boerhaave is continually discussed both as scientist and student of theology, and toward the conclusion Johnson remarks, as in the *Life of Sydenham*, that his piety is more important than his scientific attainments:

> But his Knowledge, however uncommon, holds, in his Character, but the second Place; his Virtue was yet much more uncommon than his Learning. He was an admirable Example of Temperance, Fortitude, Humility, and Devotion. His Piety, and a religious Sense of his Dependence on God, was the Basis of all his Virtues, and the Principle of his whole Conduct.[50]

Again, as in the *Life of Sydenham*, the pious scientist should be a model for all future students:

> So far was this Man from being made impious by Philosophy, or vain by Knowledge, or by Virtue, that he ascribed all his Abilities to the Bounty, and all his Goodness to the Grace of God. May his Example extend its Influence to his Admirers and Followers! May those who study his Writings, imitate his Life; and those who endeavour after his Knowledge, aspire likewise to his Piety![51]

Boerhaave's science as well as his personal character are, for Johnson, patterns of excellence, but his piety, evident in his private life as well as in the way he approached science, is of chief importance.

48. Ibid.
49. Similarly, in the *Life of Morin*, Johnson chooses two characteristics, in this case generosity and strict physical regimen, and stresses them repeatedly as he develops his biographical sketch.
50. "Boerhaave," *Med. Dict.*, I, sig. 9X1r.
51. Ibid.

The rapprochement between science and religion is of great consequence in the course of eighteenth-century thought and its importance within the ideological tradition of the new philosophy cannot be overstressed. It would indeed be surprising if Johnson did not add his voice to those of others who had considered the matter. But here, as, I think, with all of the principles of English scientific ideology to which he subscribes, he is not being carried along by an intellectual tradition in which he must perforce accept an entire series of tenets, doctrines, and convictions. On the contrary, rather than merely echoing such ideologists as Bacon, Hooke, Glanvill, Sprat, Ray, Derham, Bentley, Cowley, Dryden, and Newton, Johnson articulates their principles from the point of view of his own experience and, more often than not, in the context of discrete literary works. When he comments on satirists, projects, virtuosi, experiential learning, incredulity, "names" and "authority," superstition, technology, and the other subjects which we have treated, his remarks grow out of a body of concrete altercations, friendships, and experiences which form the basis of and add point to his generalized reflections.

That is particularly true of the relation Johnson perceives between science and religion, for ultimately he sees everything within a religious framework, and on the basis of his Christianity he welcomes the achievements of the new philosophy. But, to return to the matter with which we began, if the "humanist" is obsessed with moral rather than scientific subjects, and if self-knowledge is the final goal of his investigations, I would think Johnson might advise him that he need not necessarily discount the possibility that science might provide him some satisfaction. Near the close of his *Astro-Theology* (a book which Johnson consulted for and cited in the *Dictionary*), William Derham argues that we should not overvalue the world, nor "set our hearts too much upon it, or upon any of its Riches, Honours or Pleasures. For what is all our Globe but a Point, a Trifle to the Universe!"[52] He then quotes Matthew 16:26: "*What is a man profited, if he shall gain the whole World, and lose his own*

52. *Astro-Theology: Or a Demonstration of the Being and Attributes of God, from a Survey of the Heavens,* 2nd ed. (London, 1715), p. 220.

Soul?" In other words, the heavens show God's art, wisdom, and power, but they also demonstrate how little a creature is man and where his primary responsibilities lie. Star-knowledge brings self-knowledge, which is both humbling and grand. The danger lies in man's turning back on himself with fatuous satisfaction, rather than seeing the larger processes. This by no means excludes science; it places it in context. The accomplishments of the new philosophy are a legitimate source of pride in human capability, but reflections on the creature must lead to a consideration of the Creator. To the extent that the scientist is humbled he is exalted. Johnson as moralist is not concerned with the "goodness" or "evil" of science but with the uses to which man puts it, and, more importantly, the conclusions which man draws from it.

CHAPTER VI
The Utopian Fallacy

The energy and buoyancy which resulted from the revolutions in natural philosophy brought forth an array of ideologists who perceived man as standing, alternately, at the very brink of chaotic destruction or at the gate of a brave new world. The sheer force of the concept of scientific advance versus irrevocable regression has led men to consider science's role as central in the movement toward either earthly felicity (usually embodied in the form of a city) or torment and annihilation. Technology is a complex of tools, a series of extensions of human capability. When nature is checked, controlled, and properly unleashed (the metaphors are often bestial), the promised result is Utopia,[1] the immediate effect one of optimism and con-

1. I use the word in the sense of "good place" (eu-topia), not "no place" (ou-topia). See Nell Eurich, *Science in Utopia: A Mighty Design* (Cambridge, Mass., 1967), p. 275. Besides Mrs. Eurich's study I am indebted in this chapter to such standard works as Carl L. Becker, *The Heavenly City of the Eighteenth-Century Philosophers* (New Haven, 1932); J. B. Bury, *The Idea of Progress: An Inquiry into its Origin and Growth* (New York, 1932); R. S. Crane, "Anglican Apologetics and the Idea of Progress, 1699–

fidence. Unfortunately, the very extensions of human knowledge and physical faculties have a way of turning on their masters, like the creation of Frankenstein, and throwing us out of our heavenly cities, or their microcosm, the laboratory. Thus, precisely like the eighteenth century, we simultaneously fear and place our hopes on the capabilities provided by the new science. Scientific knowledge is both invigorating and disquieting; we can understand Condorcet's writing his *Sketch of a Historical Picture of the Progress of the Human Mind* while hiding from Robespierre, and at the same time embrace Pascal's terse and famous comment on the world which science had revealed: "Le silence éternel de ces espaces infinis m'effraye."

Such a state of affairs is of considerable interest to the historian of ideas or ideologies, for so volatile a concept as scientific progress and its relation to man's alleged perfectibility has a way of dividing thinkers and writers, casting them back upon their most basic conceptions of the nature of man, his environment, and his destiny, and forcing them to provide critiques, alternatives, or rationales for allegiance to a notion which may serve to illuminate both the intellectual history of their period and their personal intellectual tempers or predilections. As J. B. Bury, among others, has pointed out, the "hypothesis of man's moral and social 'perfectibility' . . . rests on much less impressive evidence [than the 'continuous progress in man's knowledge of his environment'],"[2] a matter of which Swift, to name the most obvious figure, is painfully aware, but a predicament which many of his contemporaries would consider nonexistent. Johnson's position falls between extremes. He welcomes technology, utilitarian applications of science, and sees such phenomena as pal-

---

1745," reprinted in *The Idea of the Humanities and other Essays Critical and Historical*, I (Chicago, 1967), 214–87; Ernest Lee Tuveson, *Millennium and Utopia: A Study in the Background of the Idea of Progress* (Berkeley and Los Angeles, 1949); and Lois Whitney, *Primitivism and the Idea of Progress in English Popular Literature of the Eighteenth Century* (Baltimore, 1934). Of particular interest is the recent *Utopias and Utopian Thought*, ed. Frank E. Manuel (Boston, 1967), especially Northrop Frye, "Varieties of Literary Utopias," pp. 25–49, and Manuel, "Toward a Psychological History of Utopias," pp. 69–98.

2. Bury, *The Idea of Progress*, p. 4.

pable, worthy advances. He realizes that the rapidity of such developments may vary and that the tedious process of attaining truth may be accelerated by the addition of new investigators. He does not, however, expect an imminent utopia; he would deny that a utopia established on secular principles could satisfy man, and he shares Swift's reservations concerning human perfectibility.

Before raising so many potentially confusing issues as eighteenth-century notions of progress, primitivism, utopia, optimism, and the millennium, it should be clearly noted that I do not intend to parade that series of spectres in full dress and then neatly retire them once and for all. Rather, I am concerned with Johnson's view of secular utopias, many, but not all of which are inspired by the successes in science or the prophetic wishes and hopes of its ideologists. Since Johnson is unsympathetic to the plans and programs of most utopists and offers a Christian alternative to their secular visions, my interest once again is in the relation, in Johnson's eyes, between science and religion. This chapter is then a logical successor to its predecessor, though its focus is different.

Such notions as progress and primitivism, optimism and utopia, can only be separated and segregated by imposing simple schemata on complex intellectual traditions and currents. A banner bearing a title such as "pastoral utopist" is erected and a series of individuals are quickly placed in line behind it. This sort of procedure may be useful and is surely justified in the case of imitators and third-line commentators who merely reflect simplified versions of common notions or preconceptions. With important writers, such a procedure—lacking extensive qualification or explanation—simply will not serve. The uniqueness of the major writer or thinker attracts us, not the fact that he could be neatly characterized or categorized. Rousseau is finally not a primitivist, but Rousseau. Swift is not a latter-day Renaissance humanist; he is Swift. However, if a figure willingly seeks to be included under a specific category, if he aspires to be worthy of a particular title, we may assume that that category or title is of more than heuristic value. Hence we are reminded time and again, quite justly I believe, that Johnson and Swift—like so many of their important contemporaries—are Christians, a fact which we are fond of forgetting, and one which considerably clarifies and illuminates their work. One may question the value of catch phrases and labels yet steadfastly affix the title "Christian" to Johnson. The title is a unique one. It is inconceivable that a person would awake at some moment in adolescence or early manhood and resolve to become the greatest "pessimist," "primitivist," or "neoclassicist" of his age, while it is quite unremarkable that a man such as Johnson should, at some moment in his life, resolve to be the most perfect Christian possible. Of course, there are many figures who have seen a quite secular mission, goal, or category in religious terms and pursued it with equal fervor. My point is that a major figure would no doubt be alarmed to see himself listed in a modern university textbook, along with a handful of his contemporaries, under some simplistic category, whereas the label "saint" or a reasonable equivalent might prove extremely appealing.

Frank E. Manuel has described the English and continental utopias between Thomas More's and the end of the eighteenth century as characterized by calm felicity, communities where monopolies of property are abolished and sexual competition diminished, agrarian, egalitarian states in which the stress falls upon serenity rather than liberty. "The mood of the system is sameness, the tonus one of Stoic calm, without excitation."[3] Little evidence need be marshalled to demonstrate Johnson's opposition to such dreams or proposals. The Epicurean valley in *Rasselas*, from which evils are allegedly excluded, begins to cloy the moment man contemplates his humanity. Rasselas' thirst and hunger may be satisfied, but his restlessness continues. The distinction which separates him from the animals and beasts, who do experience satiation, is the source of his unhappiness and his humanity. The lesson is that of Carlyle's famous shoeblack, who has a soul in addition to a stomach and whose happiness the moneyed interests of Europe cannot guarantee beyond an hour or two. The situation can be portrayed in both psychological and religious terms. In *Rambler* 103 Johnson writes:

3. Manuel, "Toward a Psychological History of Utopias," p. 76.

All the attainments possible in our present state are evidently inadequate to our capacities of enjoyment; conquest serves no purpose but that of kindling ambition, discovery has no effect but of raising expectation; the gratification of one desire encourages another, and after all our labours, studies, and enquiries, we are continually at the same distance from the completion of our schemes, have still some wish importunate to be satisfied, and some faculty restless and turbulent for want of employment.[4]

The qualification "in our present state" is central. Rasselas' aged tutor points out that in the Happy Valley all has been provided which the "emperour of Abissinia can bestow," but the satisfaction of many types of restlessness—including the Christian's—is beyond the power of Abyssinian emperors. It may be argued that Johnson's Augustinian view of human nature precludes any possibility of a utopia: perfect harmony, equality, and benevolence are not to be expected from fallen man.[5] However, the possibility of a utopia founded on secular principles is really irrelevant, for even if it were to be established, a Christian such as Johnson would realize its inadequacy. If man could construct happy valleys, the only individual likely to enjoy such an environment is, as Manuel argues, the Epicurean-Stoic. The Christian would be totally dissatisfied, for his expanded vision, made possible by revelation, balks at the limitations and impossibilities of the view of stoicism.

The secular or pagan utopias do not constitute the entire collection of perfect commonwealths and communities in the Renaissance and eighteenth century. Ernest Tuveson has shown that the doctrine of progress was linked with the concept of the

4. *Rambler*, IV, 184–85. Cf. Chester F. Chapin, *The Religious Thought of Samuel Johnson* (Ann Arbor, 1968), p. 140: "The causes of man's unhappiness are rooted in his human nature, not in his environment. Since in Johnson's view there is relatively little that social or political change can do to mitigate this unhappiness, it follows that he looks elsewhere for a cure, insofar as a cure is possible. And he finds it in the Christian hope of a better world than this." See also Walter Jackson Bate's discussion, *The Achievement of Samuel Johnson* (New York, 1955), Chapter two, "The Hunger of Imagination," pp. 63–91.

5. This is Arieh Sachs's line of argument in his *Passionate Intelligence: Imagination and Reason in the Work of Samuel Johnson* (Baltimore, 1967), Chapter six, "The Folly of Utopia," pp. 91–108.

millennium, that it was felt by some that certain books of the Bible foretell the steady rise of mankind, and that this ascent would be accompanied by an amelioration of the cultural and natural environment. In contrast to the Renaissance historiographers who thought advancement possible, millennialists such as Edmund Law see advancement as inevitable. It is, after all, revealed truth: the new philosophy, the idea of progress, and the Christian revelation are inseparably welded. Johnson, however, does not expect a "heavenly city of the virtuosi," the true utopia of the millennium, and says, in his *Dictionary* definition of the term that the concept of the millennium is "an ancient tradition in the church, grounded on a doubtful text in the Apocalypse...."

Though Johnson questions this path to a Christian utopia, he does not hesitate to offer an alternative. The text for his homily in his fifth sermon (1825 *Works*, IX, 331–41) is Nehemiah 9: 33: "Howbeit thou art just in all that is brought upon us, for thou hast done right, but we have done wickedly." Johnson's subject is the misery of life and he is interested in apportioning responsibility and blame for that misery. He quite expectedly takes to task those who "murmur at the laws of Divine providence" (p. 331) and points out that some "have endeavoured to demonstrate, *and have in reality demonstrated* to all those who will steal a few moments from noise and show, and luxury, to attend to reason and to truth, that *nothing* is worthy of our ardent wishes, or intense solicitude, that terminates in *this* state of existence, and that those only make the true use of life that employ it in obtaining the favour of God, and securing *everlasting* happiness" (p. 331, [my italics])—the lesson of *Rasselas* and *The Vanity of Human Wishes.*

Johnson argues that few of the evils of life can be justly ascribed to God, and that "a general piety" might do much to alleviate such evils. He discusses the alleged "tranquillity and satisfaction diffused through the inhabitants of uncultivated and savage countries," portraying the conventional primitivist utopia with interspersed commentary:

It is found happy to be free from contention, though that exemption be obtained, by having nothing to contend for; and an equality of condition, though that condition be far from eligible, conduces more to the peace of society, than an established and legal subordination, in which every man is perpetually endeavouring to exalt himself to the rank above him, though by degrading others, already in possession of it, and every man exerting his efforts, to hinder his inferiours from rising to the level with himself. It appears, that it is better to have no property, than to be in perpetual apprehensions of fraudulent artifices, or open invasions; and that the security arising from a regular administration of government, is not equal to that which is produced by the absence of ambition, envy, or discontent (p. 337).

The savages are indeed "in a great measure free from those [hardships and distresses] which men bring upon one another," but Johnson is not led to embrace a mode of life whose "quiet is the effect of . . . ignorance" (p. 338). His alternative may be termed an earthly, Christian utopia, one which is not millennial, but possible this very moment, "a community, in which virtue should generally prevail, of which every member should fear God with his whole heart, and love his neighbour as himself, where every man should labour to make himself 'perfect, even as his Father which is in heaven is perfect,' and endeavour, with his utmost diligence, to imitate the divine justice, and benevolence . . ." (pp. 337–38).

Having described the foundation of such a society, he examines the certain results of a general piety. One need quote only a few lines to indicate the tone and tenor of his argument: "Every man would assist his neighbour, because he would be certain of receiving assistance, if he should himself be attacked by necessity. Every man would endeavour after merit, because merit would always be rewarded. Every tie of friendship and relation would add to happiness, because it would not be subject to be broken by envy, rivalship, or suspicion. Children would honour their parents, because all parents would be virtuous; all parents would love their children, because all children would be obedient" (p. 338). Any community may arrive at this state "by the general practice of the duties of religion," and granting that happiness such as this is within human power, Johnson asks how man can conceivably accuse Providence of cruelty or negligence, considering also that this state is a mere preparation for one of much greater happiness.

By this point one is likely to say, with Nekayah, that "Whether perfect happiness would be procured by perfect goodness . . . this world will never afford an opportunity of deciding" (*Rasselas*, pp. 65–66). However, the question is one of possibility, not likelihood. Anticipating an objection such as Nekayah's, Johnson intensifies the situation at the precise moment when credibility is most likely to break at the strain:

> Let no man charge this prospect of things, with being a train of airy phantoms; a visionary scene, with which a gay imagination may be amused in solitude and ease, but which the first survey of the world will show him to be nothing more than a pleasing delusion. Nothing has been mentioned which would not certainly be produced in any nation by a general piety. To effect all this, no miracle is required; men need only unite their endeavours, and exert those abilities which God has conferred upon them, in conformity to the laws of religion (pp. 339–40).

The final tour de force is easily predicted. We are not certain that all will concur in this practice of virtue. Should all decide to agree on the matter, the time of their choosing remains unknown, but surely we as individuals—seeing the promise held out to us and realizing what is in our power—cannot delay a single moment: "An universal reformation must be begun somewhere, and every man ought to be ambitious of being the first. He that does not promote it, retards it . . ." (p. 340). This movement from primitivist theorizing to the individual charge is hortatory rhetoric of a very high order. To be sure, Johnson does not expect an immediate reformation to result from his or anyone else's sermons, but the claim that "a general piety" is very unlikely serves to perpetuate the general impiety of the present. Hence Johnson is uncompromising with regard to such a reformation's possibility, and his challenge to the individual to show some reason why the process should be delayed is, like the challenges presented Chesterfield, Macpherson, and the American slave drivers, unanswerable. No one who has looked at the world of men through Johnson's eyes would expect an immediate alteration of human action and sudden desire on the part of all to amend their lives, restructure their values, and pursue a life of Christian virtue, but by the same token, no one who has studied Johnson at any length would expect him to attempt to purvey a secular utopia, much less to hold out much hope for a secular utopia's feasibility or success. Christianity, in the fullest sense of the term, is not likely to become a reality which pervades and vitalizes Johnson's society, but if a man is to have an ideal for his society and an imperative which will direct his own life, this is assuredly Johnson's choice.

In *The Vanity of Human Wishes* Johnson articulates the limitations of purely *human* (the title's operative word) perception with the intent of demonstrating the fulness of vision which is the Christian's possession.[6] As long as one dwells in the world delimited by China and Peru, not, it should be stressed, in the enlarged Miltonic universe, human hopes will be vain: a mark of simultaneous pride and futility. In such a limited realm man must perforce "tread the dreary paths without a guide"; there is no affable archangel, Michael, or Son of God, but merely Reason, the pagan's sole standard, which pales before the claims of riches, beauty, longevity, power, fame, learning, and ambition. Swedish Charles, whose stoicism should constitute proof against the allurements of such a world, whose adventuring captures the attention of that world's falsified structure of values, is, like his fellows, reduced to a paltry exemplum. An impoverished world such as this is aptly depicted as a fatalistic labyrinth (l. 6), a deterministic, intractable, lifeless scene which Johnson's century, like the preceding period, often associates with the pagan, atomistic philosophers. It is the universe of Milton's Satanic company in which will and energy are somehow both all powerful and impotent. Johnson's alternative is not far to seek. The consonance of the human and divine wills made possible by the all-important admixture of salvific revelation contrasts with the pitiful fatalism of the pagan with no diminishing of force or certainty. Human goods are ordained by heavenly

6. I am here indebted to Howard D. Weinbrot's discussion of the poem in his *The Formal Strain: Studies in Augustan Imitation and Satire* (Chicago, 1969), pp. 193–217.

laws and the capability of attaining them is, in no uncertain terms, guaranteed.

The judgment of *Rasselas* is identical.[7] Imlac's interpolated comments on the importance and omnipresence of divine Providence remain scattered and unstressed as Johnson's philosophic travelers depart from and yet still project earthly—and hence, limited—utopias. The pagan alternatives of the sensual life, withdrawal, pastoral felicity, stoicism, and unintelligible rationalism are systematically rejected as Johnson's parable moves to its culmination at the catacombs, the end of purely *human* existence, where the intolerable burden of paganism and the joy provided by revelation are finally recognized. Rasselas' judgment is crucial: " How gloomy would be these mansions of the dead to him who did not know that he shall never die; that what now acts shall continue its agency, and what now thinks shall think on for ever." Nekayah's decision to depreciate the importance of a choice of life and choose instead to contemplate the alternatives offered by eternity, is fit and proper. Hence, nothing can be ended in the final chapter, since the travelers' earthly exis-

7. On *Rasselas* and Johnson's interpretation of Ecclesiastes see Thomas R. Preston, "The Biblical Context of Johnson's *Rasselas*," *PMLA*, 84 (March 1969), 274–81. In my judgment, Preston's point (p. 281) "that [in *Rasselas*] this life is only the temporal phase of an eternal process" is of much more importance than his claim (p. 279) that, in Johnson's view, "once man realizes his inability to find perfect happiness in this world, he can and should enjoy to the fullest the limited joys it offers." I am not trying to minimize Johnson's obvious zest and vitality, but rather point to an explicit hierarchy of values. Taking care of one's stomach—with which Johnson is certainly concerned—is of far less moment than seeing to the condition of one's soul. It seems to me that *Rasselas* indicates the shortcomings of a world devoid of heaven and the overwhelming importance of divine revelation. The pursuit of the joys of this world are hardly forbidden; they are merely unsatisfactory. And the individual who possesses the all-important truth but little else, should be envied—though not left to suffer from material deprivation—by those who possess the goods of this world. The now famous Times Square derelict who announced to Tinker, Osgood, and A. Edward Newton that "*Rasselas* is the greatest book in the world," may well have perceived the point and was not so forlorn as he appeared. See Frederick W. Hilles, "*Rasselas*, An 'Uninstructive Tale,'" in *Johnson, Boswell and their Circle* (Oxford, 1965), pp. 119–20.

tence has not terminated. To provide a rounded conclusion would be to contradict the central lesson of the penultimate chapter. Chapter XLIX is merely the earthly conclusion in which, Johnson points out with crystalline clarity, nothing is concluded. The hermit's pronouncement that "To him that lives well . . . every form of life is good" is fully vindicated.

Nonetheless, *Rasselas* and *The Vanity of Human Wishes* remain incomplete statements of the nature of Christian duty. In the greatest of Johnson's poems, the Christian is portrayed as a suppliant, prayerfully seeking and embracing the message of revelation. In *Rasselas*, the travelers' restlessness—which Johnson portrays in physical, psychological, and theological terms— is relieved but not, on theological grounds, totally removed. They receive the Christian message, but, as in *The Vanity of Human Wishes*, we do not see that message put into practice within the world of men. Johnson was fully aware of the dangerous possibilities of apathy and complacency following as a result of the otherworldly perspective. Once man is assured of the existence of a final, supreme blessing, he may, unfortunately, fill his time with impatient waiting rather than the performance of all that is included in the norm of Christian charity. Thus, Imlac's allegedly open-minded judgment of the monastic ideal in Chapter XLVII is clearly one-sided. His statement that "He that lives well in the world is better than he that lives well in a monastery" could easily be attributed to Johnson himself, whose personal comment, "I find my vocation is rather to active life"[8] must be generalized; Johnson's masterful thesis statement in this regard, to which we have already alluded, must now be treated in detail, namely the elegy on Dr. Robert Levet.

The Levet poem finds its place in that critical limbo to which many of Johnson's "minor" works have been consigned. It is periodically inspected with the intent of demonstrating Johnson's poetic facility and perennially offered in evidence as an elegiac alternative to Milton's *Lycidas*.[9] If one of my assump-

8. *Tour*, p. 42. J. P. Hardy provides a series of references to Johnson's attitudes toward monasticism. See *Rasselas*, p. 178.
9. Two discussions of the poem to which I am indebted are Bertrand H. Bronson, "Personification Reconsidered," in *New Light on Dr. Johnson*,

tions in these pages is warranted, namely, that an important facet of Johnson's contribution to the tradition of scientific ideology lies in his rhetorical skill and his ability to fuse intense, personal experience with the norms of that tradition, it is fitting to select this work for closer attention. Having indicated in the biographies of Boerhaave, Browne, and Sydenham that the study of science may enhance the spiritual life of the individual and buttress religion rather than replace it, he portrays, in the case of Levet, the demeanor and reward of a humble Christian physician as his single gift is utilized to the fullest and employed to lessen the sufferings of his fellow men.

The narrator of the Levet elegy is the Johnson who subjects his personal religious life to such searching criticism in the *Diaries, Prayers, and Annals*, here including himself with the "we," who are unable to meet the Christian standard which Levet exemplifies. Our world is a dungeon-like mine, which deludes purely human hopes, to which we, like criminals, are condemned. Our oppressive day-to-day toil is contrasted with the yearly pattern of Levet's labors and we, unlike Levet, are subjected to deaths appropriate to those whose lives are passed in mines—the sudden blasts of capricious explosions and the gradual but inevitable deaths of "slow decline." Levet's life is one of test and tempering; he is "well tried" in the scriptural sense and as he is devoted to the least of Christ's brethren ("Of ev'ry friendless name the friend") he is not, the elegist makes clear, found wanting. He fills "affection's eye" both as a memory and a tear; his wisdom is obscure, his kindness coarse, like the very atmosphere and material of the mine in which his work is carried on. His merit, appropriately, is "unrefin'd," like that material which the miner seeks; there are no primroses, pansies, violets, and cowslips in the subterranean world which he inhabits. Levet's single talent—Johnson's alteration of Matthew 25:14–30 has already been noted—is employed to the greatest extent which Providence can require, and his reward is forthcoming. His death is neither devastatingly sudden nor painfully

---

ed. Frederick W. Hilles (New Haven, 1959), pp. 189–231; Donald J. Greene, " 'Pictures to the Mind': Johnson and Imagery," in *Johnson, Boswell and their Circle*, p. 148.

gradual—the possibilities offered in the opening stanza—but effortless. His selflessness constitutes an alternative to the practices
of those who choose delusion within the mine that is this life;
his exercise of Christian charity contrasts with the nurturing of
limited human hopes which is portrayed in detail in *The Vanity
of Human Wishes*. The personal dimension in the poem, it need
hardly be added, is of compelling importance for Johnson, who
aspires to Levet's reward but who, with full awareness of the
multiplicity of talents which he has received, is intensely concerned with the success with which he has fulfilled his heavenly
charge, one of the primary sources of self-deprecation in his
devotional writings, one motivated by serious theological principle, and not, as some would have it, by personal psychological
debility.

Perhaps the most important parallel to the Levet elegy is
Michael's description of the Lazar-house in Book xi of *Paradise
Lost*.[10] Levet's "useful care" is to be found in the darkest caverns
of misery; for Milton, Death inhabits a "grim Cave" to which
we all must journey. Some die "by violent stroke" (l. 471);
others waste away through intemperance and endure such slow
declines as melancholy, atrophy, marasmus, and dropsy. As
would be expected, in both Johnson and Milton, Death hovers
above the beds of the sick; he hesitates to strike in Milton; Johnson's Levet prevents the final blow. The inhabitants of the Lazar-
house, like those in the mine, delude themselves with hope—in
Milton the delusion involves the belief that death is man's "chief
good" and Death is invoked (ll. 492–93) in pitiful, idolatrous
fashion.

The vision brings tears to Adam's quite human eyes (as, in-
cidentally, the Levet poem did to Johnson's) and when he painfully inquires as to the possibility of alternatives, Michael
counsels temperance with the promise of an effortless death.
While Johnson's image is mechanical, Michael's is one of harvest:

10. Parallels for the Lazar-house passage itself have of course been found
in such sources as *Piers Plowman* and Du Bartas. For Johnson's praise of
the passage, see *Rambler*, iv, no. 78, 47. All references to *Paradise Lost*
are to the edition of Merritt Y. Hughes (New York, 1962).

> So may'st thou live, till like ripe Fruit thou drop
> Into thy Mother's lap, or be with ease
> Gather'd, not harshly pluckt, for death mature ....
>   (ll. 535–37)

However, the wages and sufferings of old age must attend such a process, so that Michael's final admonition involves selflessness and duty: "Nor love thy Life, nor hate; but what thou liv'st/ Live well, how long or short permit to Heav'n ..." (ll. 553–54). The consolation is perhaps not so satisfying as Johnson's, for Michael's stress falls upon pain and frailty to a greater extent than Johnson's, but then, of course, the fullness of Michael's message has not yet been delivered. The journey into the world and the promise of the paradise within, which follow the depiction of the Incarnation, have not yet been accomplished; in Johnson they are *données*. One need look no further than the second paragraph of Law's *Serious Call* for Johnson's considered judgment: "He, therefore, is the devout man, who lives no longer to his own will, or the way and spirit of the world, but to the sole will of God; who considers God in everything, who serves God in everything, who makes all the parts of his common life parts of piety, by doing everything in the Name of God, and under such rules as are conformable to His glory." This is the way of Level; Johnson advises that it be ours, and prays that it may be his own.[11]

We are not nearly so far afield at this point as it may seem. Toward the end of the *Principia* Newton claims that "to discourse of [God] from the appearances of things, does certainly belong to Natural Philosophy"[12] and at the conclusion of his *Opticks* argues that the perfecting of natural philosophy will enlarge the boundaries of moral philosophy: "For so far as we can know by natural Philosophy what is the first Cause, what

11. A similar treatment of the ideal may be found in Johnson's "Christianus Perfectus" (*Samuel Johnson: Poems*, ed. E. L. McAdam, Jr., with George Milne [New Haven, 1964], pp. 344–45), for which Maurice Quinlan provides a translation; see *Samuel Johnson: A Layman's Religion* (Madison, 1964), pp. 11–12.

12. The final sentence of the penultimate paragraph in the Motte translation of 1729.

Power he has over us, and what Benefits we receive from him, so far our Duty towards him, as well as that towards one another, will appear to us by the Light of Nature."[13] The precedents in the ideological tradition for moving from a consideration of natural phenomena, or the students of such phenomena, to an examination of our moral relationships with one another are clear, and there is little difficulty to pointing out the number of Newton's would-be successors seeking to apply his scientific notions to the realm of morality or to achieve a position in that realm tantamount to his eminence in mathematical physics. Johnson may be skeptical concerning attempts to understand and explain human action and motivation with the precision of the physical sciences[14]—he continually points out the irrationality, unpredictability, and capriciousness of human choice—but there can be no doubt that he approves of the rigorous study of human psychology and moral action. After all, he is attempting to provide the results of his own experiential study of these matters in the majority of his works, while his choice of literary rather than purely discursive forms implies that in the realm of human life and morality, one must suggest, evoke, portray, and attempt to convey tone, texture, and nuance, rather than seek algebraic formulae.

The focus on utilitarian applications of the new science as a mode of Christian charity has already been treated and it is clear that the English scientific ideologists would approve of Johnson's portrait of Levet and see it as a vindication of their own orientation, what might be termed Christian empiricism or experimentalism. However, the religious framework for scientific study does much more than provide a ready justification or rationalization for certain activities. It is quite likely that many of the new philosophy's apologists were, finally, far more concerned with science than religion, and merely used the religious framework when it seemed rhetorically convenient. This is not

13. The penultimate sentence in the 4th edition (London, 1730).

14. Such attempts are not, of course, restricted to our own period. Johnson was certainly familiar with examples of the phenomenon. See Louis I. Bredvold, "The Invention of the Ethical Calculus," *The Seventeenth Century*, (Stanford, 1951), pp. 165–80.

the case with Johnson, for whom the framework is of paramount importance, a situation which systematically affects his response to a series of important issues.

An encompassing religious framework reduces the importance of all else; the larger perspective shifts attention from the minute or trivial and returns us to the commanding issue. The transient and secular points of contention from which arise heat and controversy are in a sense beneath the attention of the man whose interest is focused on Providence and eternity, for whom the one thing needful is God, the central event of human history the Incarnation, and for whom man's destiny is either eternal reward or everlasting damnation. This path has led some to a kind of immobility, which may take such forms as withdrawal, anti-intellectualism, or monasticism. This is not the way of Johnson. The religious framework becomes a healthy corrective to excess and worldly preoccupations, but does not separate him from his world. It enables him to resist the petty cavils and avoid the numberless controversies which are always there to detain the ideologist.

The framework provides considerable balance. Johnson is neither prey to the dreams and promises of the utopist or chiliast, nor does he attempt to inhabit the "humanist's" past, lamenting the passing of ages of gold, awaiting the coming of chaos and darkness in the wake of fiendish modernism. He is not tempted to prophesy or portray a secular utopia in which the sudden burgeoning of learning and fundamental alteration of the human condition is made possible by the Moderns' methodology, nor does he feel the need to sketch a dystopia in which the new philosophy has brought a new variety of hell or a gallery of bumbling fools, sitting in rapt attention with their imbecilic inventions. His is neither the self-assured smile, the heady, breathless dream, nor the sickness unto death, but rather the measured view.

Johnson would oppose any attempt to demonstrate an essential difference between the scientific and "humanist" temperament. It could be argued that the very nature of eighteenth-century science and its students precludes the possibility of the "two cultures," that intellectual ambidexterity is the rule and not the exception, but though there are numerous examples such as Newton, d'Alembert, and Goethe to support such a contention, the notion would give little comfort to Sprat as he constructed his apologia or to the Royal Society as the satirists continued their wholesale assaults. Johnson's notion of genius erases the artificial distinctions which are perpetuated, he claims, by vanity. "The true Genius is a mind of large general powers, accidentally determined to some particular direction."[15] Johnson could have written legal briefs instead of *The Vanity of Human Wishes*; "had Sir Isaac Newton applied to poetry, he would have made a very fine epic poem" (*Tour*, p. 20). The belief in limited, particularized genius is strongly attacked in *Rambler* 25:

> But of all the bugbears by which the *Infantes barbati*, boys both young and old, have been hitherto frighted from digressing into new tracts of learning, none has been more mischievously efficacious than an opinion that every kind of knowledge requires a peculiar genius, or mental constitution, framed for the reception of some ideas, and the exclusion of others; and that to him whose genius is not adapted to the study which he prosecutes, all labour shall be vain and fruitless, vain as an endeavour to mingle oil and water, or, in the language of chemistry, to amalgamate bodies of heterogeneous principles (*Rambler*, III, 138–39).

The matter may also be treated in religious terms. Northrop Frye has written, wisely I believe, that the real problem of the "two cultures" is "not the humanist's ignorance of science or vice versa, but the ignorance of both humanist and scientist about the society of which they are both citizens."[16] Johnson would, I presume, agree, but he would also place the situation in a larger context, note the fact that the scientist and "humanist" will both be judged by their Creator, and then advise them to prepare themselves, through their actions, for such an event, one which is of far greater importance than the phenomenon which C. P. Snow laments, the limited communication between Greenwich Village and the Massachusetts Institute of Tech-

15. *Lives of the Poets*, I (*Life of Cowley*), 2.
16. Frye, "Varieties of Literary Utopias," p. 32.

nology. It should be added, however, that Johnson's personal intellectual standards might require all "humanists" to be able to describe the Second Law of Thermodynamics as well as to demonstrate a knowledge of Shakespeare, and though he has been labeled in various ways and enlisted by many to reinforce their own points of view, his appreciation of and fascination with the workings of machinery, would prohibit his being termed a natural Luddite.

Johnson's relationship with the continental enlightenment has not yet received the attention it merits, though Robert Shackleton has provided an important preliminary sketch.[17] His few oral and written statements leave one with the initial impression that his judgment of the *philosophes* is consistently hostile. During his eight-week visit to France with Baretti and the Thrales in 1775, he saw, Shackleton writes, "not a single leading figure of the Enlightenment, and no one who could lay undisputed claim to the title *philosophe*. He met some figures of the other camp, notably Fréron, the virulent enemy of Voltaire and editor of the *Année littéraire*."[18] However, the initial impression is a misleading one. Johnson's agreement with the methodology of the new philosophy, sympathy with Lockean principles of epistemology, alignment with modernism, and skeptical temper are major points of contact with the *philosophes*. Voltaire would welcome his depreciation of Leibniz and, with his fellows, Johnson's affection for Bayle. There are several interesting interactions, such as the fact that the *Encyclopédie* article *Anglois* is largely taken from the history of English in the preface to Johnson's *Dictionary*. Diderot, with his colleagues Toussaint and Eidous, translated James's *Medicinal Dictionary*, a project which involved the translation into French of such works as Johnson's *Life of Tournefort*, itself translated from the French of Fontenelle. There are many details which deserve commentary, such as Johnson's possession of the works of the Benedictine monk, Benito Feijóo, an important figure in the

17. Shackleton, "Johnson and the Enlightenment," in *Johnson, Boswell and their Circle*, pp. 76–92.
18. Ibid., pp. 77–78.

enlightenment in Spain,[19] or the common sources of the *Dictionary* and *Encyclopédie*.

Nevertheless, a study of Johnson's connections with and attitudes toward the continental enlightenment, besides pointing up a series of areas of close agreement, would of course reveal one fundamental division, namely Johnson's insistence that the business of enlightenment—the routing of superstition, growth of technology, jettisoning of various hampering forms of authority, the general spread of the scientific spirit—be carried on within a Christian framework. It is the materialism, the atheism, the paganism of the enlightenment which triggers Johnson's outrage. To adopt Peter Gay's phrase, he could never, like the *philosophes*, seek "Newton's Physics Without Newton's God."[20] This issue is of sufficient importance, in Johnson's eyes, to damn the enlightenment, even though the holy trinity of Bacon, Locke, and Newton, and the philosophical and methodological underpinnings of the enlightenment (especially as imported by Voltaire and the Dutch physicians) earn his constant regard. He is at one with the *philosophes* on a vast number of matters, but totally separated from them on the one which in his eyes is most important of all.

This rigid adherence to the principle that scientific investigation must be conducted within a religious context is neither aberrational nor surprising. The Augustinian focus on the condition of postlapsarian man, which Johnson quite expectedly dwells upon, finds parallel principles in the tradition of English scientific ideology. Utopists might find inspiration in the successes of the new philosophy and envision the hasty alleviation of all difficulties, the throwing off of all shackles, but Bacon had continually pointed out the fallibility of human perception and the fact that experiments need to be arranged, controlled, and directed in such a way that there will be compensation for the weaknesses of man's senses. Locke, who was both influenced

19. *A Catalogue of the Valuable Library of Books of the Late Learned Samuel Johnson, Esq; LL.D.* (London, 1785), items 494, 547. Johnson owned the Spanish edition as well as Brett's English translation. Also, see above pp. 64–65.

20. *The Enlightenment: An Interpretation*, II (New York, 1969), 140–50.

by and influential for the methodology of the scientists, continually seeks to define the limitations of human knowledge. He does not share Descartes' confidence in the powers of reason, attempts to perceive not only the nature and origin but also the extent of our knowledge, and tries to restrain the exaggerated claims and pretenses concerning human knowledge of which so many are fond. One could realize the eighteenth-century's awareness of human limitation in a fallen world from the period's scientific literature as well as from its religious, and this is but one of the important notions which, in England, binds the two.[21]

However, the awareness of limitation, whether it be stated in scientific, theological, or epistemological terms, coexists with such undeniable achievements as Newton's, which tend to alter and extend the accepted notions of human capability. The tendency toward a relatively static conception of human intellectual accomplishment is further complicated by such practical matters as the increase in the number of scientific investigators and the replacement of indefatigable but limited publicists with a series of both technical and popular scientific journals. The situation must be kept in mind, for in light of it, many of Johnson's ostensibly contradictory judgments are quite consistent. He can, for example, expect little from fragile and fallible man, yet at the same time praise man's magnificent accomplishments. Depending on the passage in question, he can be discussed as the enlightenment's sternest critic or one of its most distinguished representatives. He can challenge his contemporaries and demand the alleviation of human ignorance and suffering within the confines of this world, yet speak of eternity in terms which make the secular realm shrink into in-

21. Cf. Donald Greene, *The Age of Exuberance* (New York, 1970) p. 104: "the moral and psychological basis of the 'new philosophy,' as of Augustinian Christianity, is the derogation of the inherent powers of human nature, in particular human reason. By adopting an attitude of humility as to what man can accomplish without external aid—in morality, from God; in science, from God's creation—one can learn both to love and to know. If this proposition is true, there is no need to marvel at Addison . . . hailing the discoveries of . . . Newton . . . or at the devout Samuel Johnson choosing as the epigraph for . . . *The Rambler*, the great 'skeptical' and 'anti-authoritarian' motto of the Royal Society. . . ."

significance. The variety of Johnson's responses to the new philosophy is prompted, finally, by the central ambiguities of human life, by the flexibility of the tradition of English scientific ideology, and by an individual temper which is practical, sympathetic, open, encouraging, yet—when the situation demands it—unyielding.

APPENDICES

INDEX

Johnson and the
Texts of the *Life of Boerhaave*

The importance of the Life of Boerhaave for an understanding of Johnson's attitudes toward science is undeniable, and it was surely written, as S. C. Roberts suggests, *con amore*.[1] The biography was reprinted frequently during Johnson's lifetime, and we may justly inquire whether or not he was responsible for the various reprintings. In discussing Johnson's contributions to James's *Medicinal Dictionary*, Allen T. Hazen noted that the 1739 *Gentleman's Magazine* biography was not only enlarged for that work but also reappeared in the *Universal Magazine* for February and March, 1752. Since the *Universal Magazine* edition included six paragraphs on Boerhaave's *Indexes*, which formed a part of the article "Botany" in the *Medicinal Dictionary* and was not included in the Boerhaave entry, Hazen concluded that the *Universal Magazine* text is the authentic one, that "no one but Johnson himself, or a man like Dr. James who knew the work and the authorship," would have combined the Boerhaave entry and the discussion of the *Indexes* to form a complete biography.[2] However, there is considerable varia-

1. *Doctor Johnson and Others* (Cambridge, 1958), p. 83.
2. Allen T. Hazen, "Samuel Johnson and Dr. Robert James," *Bulletin of the Institute of the History of Medicine*, 4 (June 1936), 456–57.

tion among the three texts, the nature of which suggests that if Johnson was responsible for the *Universal Magazine* edition, which is doubtful, the text is still not authoritative.

Including additions and a handful of minor deletions, there are over 85 substantive changes from the *Gentleman's Magazine* edition in the *Medicinal Dictionary*. The bibliography was completely overhauled; wholly new sections on Boerhaave's *Institutes*, *Aphorisms*, *Chemistry*, and *Indexes* were added; confused passages were revamped.[3] The *Medicinal Dictionary* text, for which Johnson was undoubtedly responsible, represents a major revision of his 1739 biography. The *Universal Magazine* edition, on the other hand, tends to increase the number of errors rather than rectify them. The size of Boerhaave's family, for example, is confused and the Cartesian Professor of Franeker, one of Boerhaave's many assailants, is ludicrously referred to as "Professor Franeker," the sort of error attributable to careless haste and a lack of acquaintance with material that Johnson had gone over on two separate occasions.[4] Of much greater importance is the fact that the *Universal Magazine* edition's more than 125 substantive changes from the *Medicinal Dictionary* version often consist of deletions. Approximately 850 words of text, or one-twelfth of the biography, is missing.[5] I would conclude

3. For example, in the *Gentleman's Magazine* (p. 175), Johnson wrote that Boerhaave "desired only to think of God, what God knows of himself." In attempting to praise Boerhaave's humility he has described instead consummate theological pride. Perhaps "what God shows of himself" was intended (a suggestion I owe to Robert Haig); at any rate the alteration in the *Medicinal Dictionary* (I, sig. 9X1^r) is explicit: "He desired only to think of God, what God has reveal'd of himself."

4. The *Universal Magazine* version does correctly alter a date (from 1726 to 1729, p. 56; *Med. Dict.*, I, sig. 9U2^v), probably a typographical slip and oversight originally.

5. Even such characteristic passages as the following (*Med. Dict.*, I, sig. 9U2^v) are omitted: "Yet I cannot but implore, with the greatest Earnestness, such as have been conversant with this great Man, that they will not so far neglect the common Interest of Mankind, as to suffer any of these Circumstances to be lost to Posterity. Men are generally idle, and ready to satisfy themselves, and intimidate the Industry of others, by calling that impossible which is only difficult. The Skill to which *Boerhaave* attained, by a long and unwearied Observation of Nature, ought therefore to be transmitted in all its Particulars to future Ages, that his Successors may be ashamed to fall below him, and that none may hereafter excuse his Ignorance, by pleading the Impossibility of clearer Knowledge."

that the *Universal Magazine* edition was a potboiling venture in which accuracy and faithfulness were sacrificed in the interest of haste and the exigencies of available space. James may well have been responsible for it, but Johnson almost surely was not. In any case, the text is sufficiently truncated and corrupt to deny it any authority.

The biography was reprinted at least one other time during Johnson's lifetime, for Davies included it in the *Miscellaneous and Fugitive Pieces*, which follows the *Medicinal Dictionary* version.[6] Moreover, *Some remarkable passages of the life and death of the celebrated Dr. Boerhaave* appeared both in the *Annual Register* (i, 245–47) and the *London Chronicle* (iv, 570) for 1758.[7] These texts are nearly identical (there are 7 minor variants in the course of the approximately 1500-word texts), and consist of a series of passages from Johnson's biography, reorganized and stitched together with commentary. They are probably based on the *Gentleman's Magazine* life, which was more accessible than the *Medicinal Dictionary* version, but it is difficult to assess textual debts since the text is freely tampered with and the majority of the passages cited do not vary in the three previous editions. Since there are references to material in Johnson's source, Schultens' *Oratio academica in memoriam Hermanni Boerhaavi*, included in the *Gentleman's Magazine* and *Medicinal Dictionary* versions but dropped in the *Universal Magazine*, it is nearly certain that the latter was not used.

Over a decade ago, F. W. Gibbs suggested that Johnson may be responsible for one of the rarest pieces in the history of chemistry, a translation from the Latin of the first eight sheets of Boerhaave's *Elementa Chemiae* (publ. 10 Jan. 1732).[8] Gibbs's argument may appear persuasive, but I doubt that the work is Johnson's. The sale catalogue of Johnson's library includes two 1732 editions of the *Elementa Chemiae*. Following Gibbs's tentative attribution one would assume that these were acquired because of a personal and immediate inter-

6. Though there are over 45 alterations of the *Med. Dict.* version in Davies' text, they are generally quite petty: change of articles, prepositions; minor reorganization of word order, pluralization, etc. The discussion of Boerhaave's *Indexes* does not appear in Davies' text.

7. Noted by Edward A. Bloom, *Samuel Johnson in Grub Street* (Providence, 1957), p. 275, n. 64, though his phrasing suggests that the entire biography was reprinted, which is not the case.

8. "Dr. Johnson's First Published Work?", *Ambix*, 8 (Feb. 1960), 24–34. Arthur Sherbo has also challenged this attribution. See his "The Translation of Boerhaave's *Elementa Chemiae*," and Gibbs's answer, ibid., 13 (June 1966), 108–117.

est in the work. I would say, on the other hand, that these were probably acquired after 1739, perhaps at the time of the work on the *Medicinal Dictionary*. In the preface to the 1732 *Elementa Chemiae* Boerhaave discusses a spurious edition of his chemistry published without his knowledge or consent, and includes a bibliography of authoritative publications. With one exception this is the list which Johnson includes in the *Medicinal Dictionary* biography (the last item—*Epistola pro Sententia Malpighiana de Glandulis ad Cl. Ruischium*—is omitted). The *Gentleman's Magazine* list, on the other hand, is briefer, less precise concerning bibliographical details, and lacks eight items which appear in the second version of the biography as well as the 1732 *Elementa Chemiae*. I would consider it highly unlikely that Johnson would have translated the eight sheets, owned two copies of the work, and suddenly forgotten the bibliography contained therein when he began to write the *Gentleman's Magazine* biography. If he had the work before him when he composed a few years later for James, why would he have overlooked it in 1739? Moreover, in listing Boerhaave's "genuine Works," Johnson alludes to the 1732 bibliography. The entire paragraph is missing in the *Gentleman's Magazine* version, and the somewhat offhand introductory paragraph to the list in the *Gentleman's Magazine* is jettisoned in the *Medicinal Dictionary*.

Johnson's admiration for Boerhaave was lifelong, but I would conclude that the only works concerning Boerhaave which can presently be attributed to him are the 1739 *Gentleman's Magazine* and the 1743 *Medicinal Dictionary* biographies.

APPENDIX B

Johnson and
the *Medicinal Dictionary*

Professor Hazen and Dr. McHenry have attributed the following
entries in James's *Medicinal Dictionary* to Johnson[1]:

"Actuarius":	I, sigs. Ii1^v–Ii2^r
"Aegineta":	I, sig. Nn1^v
"Aesculapius":	I, sigs. Uu2^v–Xx2^r
"Aetius":	I, sigs. Yy2^r–Zz1^r
"Alexander":	I, sig. Sss1^r
"Archagathus":	I, sigs. 7T1^r–7T1^v
"Aretaeus":	I, sigs. 7U1^r–7U2^v
"Asclepiades":	I, sigs. 8L2^v–8M1^v
"Boerhaave":	I, sigs. 9U1^r–9X1^v
"Oribasius":	III, sigs. Nn*2^v–Oo*1^r
"Ruysch":	I, sigs: 5Z2^r–6A1^r (under "Anatome")

1. See Allen T. Hazen, "Samuel Johnson and Dr. Robert James," *Bulletin of the Institute of the History of Medicine*, 4 (June 1936), 455–64; Hazen, "Johnson's Life of Frederic Ruysch," ibid., 7 (March 1939), 324–34; and Lawrence C. McHenry, Jr., "Dr. Samuel Johnson's Medical Biographies," *Journal of the History of Medicine and Allied Sciences*, 14 (1959), 298–310.

Historical/Biographical
Section of
"Botany":   I, sigs. 10B1ʳ–10E1ʳ (including the *Life of Tour-nefort*, treatment of Tournefort's system, and discussion of Boerhaave's *Indexes*)

The attributions have been, it is obvious, almost completely confined to biographical entries—the sort of contribution one would expect from Johnson, a contribution which could be made with little necessity for expertise and which could be accomplished quite easily through the use of secondary sources. To my knowledge, no one has attempted to sift through the 3,000-odd folio pages of technical entries in an attempt to discern Johnson's hand. First, it would be unlikely that James would call on Johnson to do something which he could well do himself more efficiently and expeditiously. Second, it would be very difficult—in many cases impossible—to find any evidence whatsoever for attributing many articles to James, much less to Johnson. A simple, bland definition, for example, or an extended quotation from Galen, could be the work of anyone. Finally, the sheer size of the work is sufficient to inhibit an undertaking which would easily consume many years and quite likely bring little or no return for one's efforts. Thus, attention has focused on the biographies, and even there the evidence supporting attributions is often very thin.

Though many of the biographies are translated or adapted from other sources and our familiarity with Johnson's methods of translation can facilitate attribution, there is little to keep Dr. James from epitomizing, reorganizing, or editorializing in the Johnsonian manner. He knows sources such as LeClerc, Photius, and Freind as well as Johnson and is perfectly capable of using material taken from them. Second, if we are to accept the work's long preface as assuredly James's then Johnson and James are in agreement concerning important ideological principles, which weakens thematic—as opposed to stylistic—internal evidence. There, James, like Johnson, attacks "fine-spun Speculations," assails obscurity and demands purity of style, and mentions, in Modern fashion, Aristotle's "Philosophical Romances."[2] His defense of the Moderns' methodology, which he associates with Hippocrates, is uncompromising.[3] A third difficulty

2. *Med. Dict.*, I, xxx; xxxi; lix; xxxv.
3. Ibid., xxxv: "The present State of Physic, and the perpetual and too successful Attempts which have been made in all Ages by Philosophers of

in attributing biographies lies in the fact that James himself is so fond of writing them. The study of a particular branch of science or medicine is, for James, a study of the history of a series of individual scientists or physicians. Under the article "Anatome," for example, Dr. McHenry has counted some 175 biographies,[4] and the pattern is hardly restricted to that article. In other words, from the hundreds of biographies—some, of course, only a few lines in length—a handful have been tentatively accepted as Johnson's; with the exception of the *Life of Boerhaave*, reliance on internal evidence is total.

The possibility of another writer's catching the flavor of Johnsonian prose in isolated passages is not remote and in some cases attributions are based on single sentences. In short, the task of attribution in James's work is extremely difficult and the work of Hazen and McHenry has been invaluable. With one exception, I am willing to accept the present attributions. I have inspected all of the biographical entries in James's work, but I consider the evidence too slight to justify further attributions at this time.

The attribution in question is the last, the historical-biographical section of the article "Botany" from sigs. 10B1ʳ–10C1ᵛ. At sig. 10C1ᵛ the *Life of Tournefort* begins, which is clearly divided, by separate title, from the biographies which precede it. The other biographies do not strike me as Johnsonian, though in many cases their preoccupation with bibliography and antiquarian detail is quite Jamesian. I see nothing in a sketch such as this, for example, to justify attributing it to Johnson:

> *Leonardus Rauvolsius* was born at *Mechlin* in 1517. He travell'd thro' *Syria, Judea, Arabia, Mesopotamia, Babylon, Assyria,* and *Armenia,* from which Countries he brought back into *Germany,* with him, many Herbs, Shrubs, Plants, and other things of the like Nature. He wrote a Book, which he calls *Hodoeporicon,* or Travels into *Syria, Judea, Arabia, Mesopotamia, Babylon, Assyria,* and *Armenia,* which he divided into six Parts, and which contain many curious things relating to the *Materia Medica.* He flourish'd about the Year 1583 (sig. 10C1ʳ).

---

all Sects, to destroy the Progress it had already made, and retard its farther Improvement, give us abundant Reason to lament, that the Scheme of *Hippocrates* was not pursued; for in all succeeding Ages, we shall have the Mortification of finding subtle Hypotheses, trifling Distinctions, whimsical, or at best uncertain Causes, and an unmeaning Jargon of Words, substituted instead of Details of Facts established by accurate Observations, and of unquestionable Events confirmed by Experience."

4. McHenry, "Dr. Samuel Johnson's Medical Biographies," p. 301.

On the other hand, when we come to the *Life of Tournefort* the opening sentence suggests a shift in emphasis which is unmistakable:

When we observe any Man distinguish'd by a superior Knowledge, or Skill of any Kind, it is natural for the Mind to be solicitous and inquisitive about the several Circumstances which have concurr'd to render him thus conspicuous.

My guess, and it is little more than that, is weakened by the fact that the treatment of Boerhaave's *Indexes* is contiguous with a brief historical/biographical section (sig. 10D2^v) following the sections on Tournefort and his system, which does, at points, sound like Johnson. This would mean that Johnson wrote one part of the short historical/biographical entries, and the separated section on Tournefort, but not the initial historical/biographical survey, a possible but somewhat clumsy method of composition.

Index